Columbia University Lectures

GRAPHICAL METHODS

ERNEST KEMPTON ADAMS RESEARCH FUND

1909-1910

COLUMBIA
UNIVERSITY PRESS
SALES AGENTS

NEW YORK:
LEMCKE & BUECHNER
30–32 WEST 27TH STREET

LONDON:
HENRY FROWDE
AMES CORNER, E.C.

TORONTO:
HENRY FROWDE
25 RICHMOND STREET, W.

COLUMBIA UNIVERSITY LECTURES

GRAPHICAL METHODS

BY

CARL RUNGE, Ph.D.

PROFESSOR OF APPLIED MATHEMATICS IN THE UNIVERSITY OF GÖTTINGEN
KAISER WILHELM PROFESSOR OF GERMAN HISTORY AND INSTITUTIONS
FOR THE YEAR 1909–1910

New York
COLUMBIA UNIVERSITY PRESS
1912

PRESS OF
THE NEW ERA PRINTING COMPANY
LANCASTER, PA.

INTRODUCTION.

§ 1. A great many if not all of the problems in mathematics may be so formulated that they consist in finding from given data the values of certain unknown quantities subject to certain conditions. We may distinguish different stages in the solution of a problem. The first stage we might say is the proof that the quantities sought for really exist, that it is possible to satisfy the given conditions or, as the case may be, the proof that it is impossible. In the latter case we have done with the problem. Take for instance the celebrated question of the squaring of the circle. We may in a more generalized form state it thus: Find the integral numbers, which are the coefficients of an algebraic equation, of which π is one of the roots. Thirty years ago Lindemann showed that integral numbers subject to these conditions do not exist and thus a problem as old almost as human history came to an end. Or to give another instance take Fermat's problem, for the solution of which the late Mr. Wolfskehl, of Darmstadt, has left \$25,000 in his will. Find the integral numbers x, y, z that satisfy the equation

$$x^n + y^n = z^n,$$

where n is an integral number greater than two. Fermat maintained that it is impossible to satisfy these conditions and he is probably right. But as yet it has not been shown. So the solution of the problem may or may not end in its first stage.

In many other cases the first stage of the solution may be so easy, that we immediately pass on to the second stage of finding methods to calculate the unknown quantities sought for. Or even if the first stage of the solution is not so easy, it may be expedient to pass on to the second stage. For if we succeed in finding methods of calculation that determine the unknown quantities, the proof of their existence is included. If on the other hand, we do not succeed, then it will be time enough to return to the first stage.

There are not a small number of men who believe the task of the mathematician to end here. This, I think, is due to the fact that the pure mathematician as a rule is not in the habit of pushing his investigation so far as to find something out about the real things of this world. He leaves that to the astronomer, to the physicist, to the engineer. These men, on the other hand, take the greatest interest in the actual numerical values that are the outcome of the mathematical methods of calculation. They have to carry out the calculation and as soon as they do so, the question arises whether they could not get at the same result in a shorter way, with less trouble. Suppose the mathematician gives them a method of calculation perfectly logical and conclusive but taking 200 years of incessant numerical work to complete. They would be justified in thinking that this is not much better than no method at all. So there arises a third stage of the solution of a mathematical problem in which the object is to develop methods for finding the result with as little trouble as possible. I maintain that this third stage is just as much a chapter of mathematics as the first two stages and it will not do to leave it to the astronomer, to the physicist, to the engineer or whoever applies mathematical methods, for this reason that these men are bent on the results and therefore they will be apt to overlook the full generality of the methods they happen to hit on, while in the hands of the mathematician the methods would be developed from a higher standpoint and their bearing on other problems in other scientific inquiries would be more likely to receive the proper attention.

The state of affairs today is such that in a number of cases the methods of the engineer or the surveyor are not known to the astronomer or the physicist, or vice versa, although their problems may be mathematically almost identical. It is particularly so with graphical methods, that have been invented for definite problems. A more general exposition makes them applicable to a vast number of cases that were originally not thought of.

In this course I shall review the graphical methods from a general standpoint, that is, I shall try to formulate and to teach them in their most generalized form so as to facilitate their application in any problem, with which they are mathematically connected.[1] The student is advised to do practical exercises. Nothing but the repeated application of the methods will give him the whole grasp of the subject. For it is not sufficient to understand the underlying ideas, it is also necessary to acquire a certain facility in applying them. You might as well try to learn piano playing only by attending concerts as to learn the graphical methods only through lectures.

[1] For the literature of the subject see "Encyklopädie der mathematischen Wissenschaften," Art. R. Mehmke, "Numerisches Rechnen," and Art. F Willers and C. Runge, "Graphische Integration."

CHAPTER I.

GRAPHICAL CALCULATION.

§ 2. *Graphical Arithmetic.*—Any quantity susceptible of mensuration can be graphically represented by a straight line, the length of the line corresponding to the value of the quantity. But this is by no means the only possible way. A quantity might also be and is sometimes graphically represented by an angle or by the length of a curved line or by the area of a square or triangle or any other figure or by the anharmonic ratio of four points in a straight line or in a variety of other ways. The representation by straight lines has some advantages over the others, mainly on account of the facility with which the elementary mathematical operations can be carried out.

What is the use of representing quantities on paper? It is a convenient way of placing them before our eye, of comparing them, of handling them. If pencil and paper were not as cheap as they are, or if to draw a line were a long and tedious undertaking, or if our eye were not as skillful and expert an assistant, graphical methods would lose much of their significance. Or, on the other hand, if electric currents or any other measurable quantities were as cheaply and conveniently produced in any desired degree and added, subtracted, multiplied and divided with equal facility, it might be profitable to use them for the representation of any other measurable quantities, not so easily produced or handled.

The addition of two positive quantities represented by straight lines of given length is effected by laying them off in the same direction, one behind the other. The direction gives each line a beginning and an end. The beginning of the second line has to coincide with the end of the first, and the resulting line representing the sum of the two runs from the beginning of the first to the end of the second.   Similarly the subtraction of one positive quantity from another is effected by giving the lines opposite directions and letting the beginning of the line that is to be subtracted coincide with the end of the other.   The result of the subtraction is represented by the line that runs from the beginning of the minuend to the end of the subtrahend.   The result is positive when this direction coincides with that of the minuend, and negative when it coincides with that of the subtrahend.   This leads to the representation of positive and negative quantities by lines of opposite direction.   The subtraction of one positive quantity from another may then be looked upon as the addition of a positive and a negative quantity.   I do not want to dwell on the logical explanation of this subject, but I want to point out the practical method used for adding a large number of positive and negative quantities represented by straight lines of opposite direction.   Take a straight edge, say a piece of paper folded over so as to form a straight edge, mark a point on it, and assign one of the two directions as the positive one.   Lay the edge in succession over the different lines and run a pointer along it through an amount equal in each case to the length of the line and in the positive or negative direction according to the sign of the quantity.   The pointer is to begin at the point marked.   The line running from this point to where the pointer stops represents the sum of the given quantities.   The advantage of this method is that the intermediate positions of the pointer need not be marked provided only that the pointer keeps its position during the movement of the edge from one line to the next.   As an example take the area, Fig. 1.   A number of

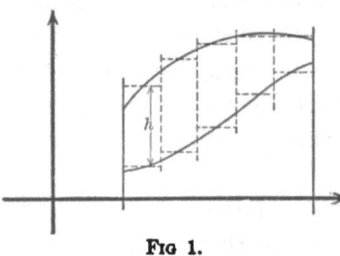

FIG 1.

rectangular strips ½ cm. wide are substituted for the area so that, measured in square centimeters, it is equal to half the sum of the lengths of the strips measured in centimeters.   The straight

edge is placed over the strips in succession and the pointer is
run along them. The edge is supposed to carry a centime-
ter scale and the pointer is to begin at zero. The final position
of the pointer gives half the value of the area in square centimeters. The drawing of the strips may be dispensed with, their
lengths being estimated, only their width must be shown. If
the scale should be too short for the whole length, the only thing
we have to do is to break any of the lengths that range over the
end of the scale and to count how many times we have gone
over the whole scale. I have found it convenient to use a little
pointer of paper fastened on the runner of a slide rule so that it
can be moved up and down the metrical scale on one side of the

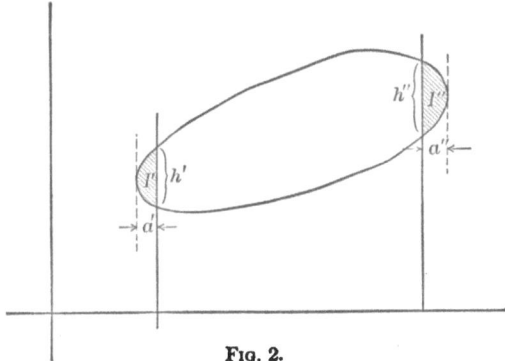

Fig. 2.

slide rule. The area is in this manner determined rapidly and
with considerable accuracy, very well comparable to the accuracy of a good planimeter. If the area of any closed curve
is to be found, the way to proceed is to choose two parallel
lines that cut off two segments on either side (see Fig. 2), to
measure the area between them by the method described above
and to estimate the two segments separately. If the curves of
the segments may with sufficient accuracy be regarded as arcs
of parabolas the area would be two thirds the product of length
and width. If not they would have to be estimated by substituting a rectangle or a number of rectangles for them.

In the same way the addition and subtraction of pure numbers may also be carried out. We need only represent the numbers by the ratios of the lengths of straight lines to a certain fixed line. The ratio of the length of the sum of the lines to the length of the fixed lines is equal to the sum of the numbers. The construction also applies to positive and negative numbers, if we represent them by the ratio of the length of straight lines of opposite directions to the length of a fixed line.

In order to multiply a given quantity c by a given number, let the number be given as the ratio of the lengths of two straight lines a/b. If the quantity c is also represented by a straight line, all we have to do is to find a straight line x whose length is to the length of c as a to b. This can be done in many ways by

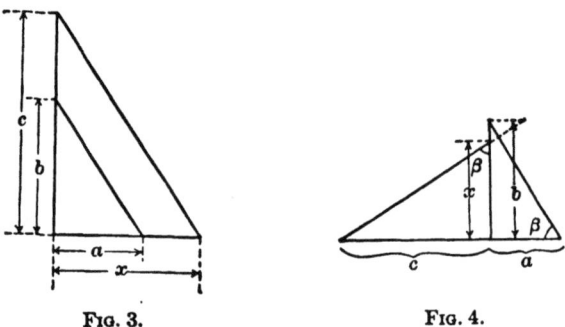

FIG. 3.   FIG. 4.

constructing any triangle with two sides equal to a and b and drawing a similar triangle with the side that corresponds to b made equal to c. As a rule it is convenient to draw a and b at right angles and the similar triangle either with its hypotenuse parallel (Fig. 3) or at right angles (Fig. 4) to the hypotenuse of the first triangle. Division by a given number is effected by the same construction; for the multiplication by the ratio a/b is equivalent to the divisions by the ratio b/a.

If a, b, c are any given numbers, we can represent them by the ratios of three straight lines to a fixed line. Then the ratio cf

the line constructed in the way shown in Fig. 3 and Fig. 4 to
the fixed line is equal to the number

$$\frac{ac}{b}.$$

Multiplication and division are in this way carried out simultaneously. In order to have multiplication alone, we need only make b equal 1 and in order to have division alone, we need only make a or c equal 1.

In order to include the multiplication and division of positive and negative numbers we can proceed in the following way. Let the lines corresponding to a, x, Fig. 3, be drawn to the right side of the vertex to signify positive numbers and to the left side to signify negative numbers. Similarly let the lines corresponding to b, c be drawn upward to signify positive numbers and downward to signify negative numbers. Then the drawing of a parallel to the hypotenuse of the rectangular triangle a, b through the end of the line corresponding to c will always lead to the number

$$x = \frac{ac}{b}$$

whatever the signs of a, b, c may be.

The same definition will not hold for the construction of Fig. 4. If the positive direction of the line corresponding to a is to the right and the positive direction of the line corresponding to b is upwards then the positive directions of x and c ought to be such that when the right-angled triangle x, c is turned through an angle of 90° to make the positive direction of x coincident with the positive direction of a, the positive direction of c coincides with the positive direction of b. If we wish to have the positive direction of x upward, the positive direction of c would have to be to the left, or if we wish to have the positive direction of c to the right, the positive direction of x would have to be downward. If this is adhered to, the construction for division and multiplication will include the signs.

§ 3. *Integral Functions.*—We have shown how to add, subtract, multiply, divide given numbers graphically by representing them as ratios of the lengths of straight lines to the length of a fixed line and finding the result of the operation as the ratio of the length of a certain line to the same fixed line. By repeating these constructions we are now enabled to find the value of any algebraical expression built up by these four operations in any succession and repetition. Let us see for instance how the values of an integral function of x, that is to say, an expression of the form

$$a_0 + a_1 x + a_2 x^2 + \cdots + a_n x^n$$

may be found by geometrical construction, where $a_0, a_1 \cdots a_n, x$

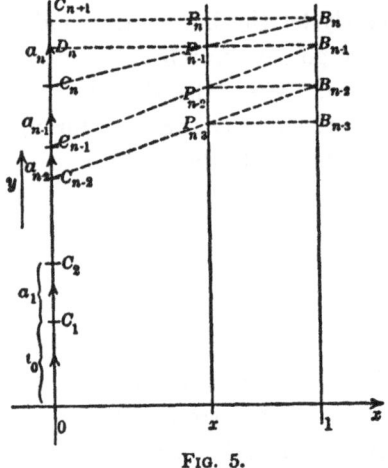

FIG. 5.

are any positive or negative numbers. We shall first assume that all the numbers are positive, but there is not the least difficulty in extending the method to the more general case.

Now let $a_0, a_1, a_2, \cdots a_n$ signify straight lines laid off on a vertical line that we call the y-axis, one after the other as if to find the straight line

$$a_0 + a_1 + a_2 + \cdots + a_n.$$

The lengths of these lines measured in a conveniently chosen unit of length are equal to the numbers designated by the same letters. In Fig. 5 a_0 runs from the point O to point C_1, a_1 from C_1 to C_2, $\cdots a_n$ from C_n to C_{n+1}.

Let x be the *ratio* of the lines Ox and $O1$, Fig. 5, drawn horizontally from O to the right. The length $O1$ is chosen of convenient size independent of the unit of length that measures the lines $a_0, a_1, \cdots a_n$. The length Ox is then defined by the value of the ratio x. Through x and 1 draw lines parallel to the y-axis. Through C_{n+1} draw a line parallel to Ox, that intersects those two parallels in P_n and B_n. Draw the line $B_n C_n$ that intersects the parallel through x in P_{n-1}. Then the height of P_{n-1} above C_n will be equal to $a_n x$. For if we draw a line through P_{n-1} parallel to Ox intersecting the y-axis in D_n, the triangle $C_n D_n P_{n-1}$ will be similar to $C_n C_{n+1} B_n$ and their ratio is equal to x, therefore $C_n D_n = a_n x$. Consequently the height of P_{n-1} above C_{n-1} is equal to $C_{n-1} D_n = a_n x + a_{n-1}$. Now let us repeat the same operation in letting the point D_n take the part of C_{n+1}. Through D_n draw a line parallel to Ox, that intersects the parallels through x and 1 in P_{n-1} and B_{n-1}. Draw the line $B_{n-1} C_{n-1}$ that intersects the parallel through x in P_{n-2}.

Then the height of P_{n-2} above C_{n-1} will be equal to

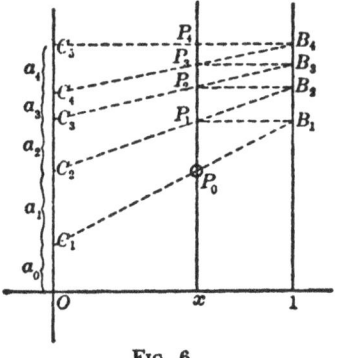

FIG. 6.

$$C_{n-1} D_n \cdot x = (a_n x + a_{n-1})x,$$

and the height above C_{n-2} will be equal to

$$a_n x^2 + a_{n-1} x + a_{n-2}.$$

Continue in the same way. Draw $P_{n-2} B_{n-2}$ parallel to Ox, draw $B_{n-2} C_{n-2}$ and find the point P_{n-3}. Then the height of P_{n-3} above C_{n-2} will be

$$(a_n x^2 + a_{n-1} x + a_{n-2})x$$

and the height of P_{n-3} above C_{n-3}

$$a_n x^3 + a_{n-1} x^2 + a_{n-2} x + a_{n-3}.$$

Finally a point P_0 is found (see Fig. 6 for $n = 4$) by the intersection of $B_1 C_1$ with the parallel to the y-axis through x, whose height above O is equal to

$$a_n x^n + a_{n-1} x^{n-1} + \cdots + a_1 x + a_0$$

Let us designate the line $x P_0$ by y, so that

$$y = a_n x^n + a_{n-1} x^{n-1} + \cdots + a_1 x + a_0,$$

in the sense that y is a vertical line of the same direction and length as the sum of the vertical lines $a_n x^n$, $a_{n-1} x^{n-1}$, $\cdots a_1 x$, a_0.

The same construction holds good for values of x greater than 1 or negative. The only difference is that the point x is beyond the interval $O1$ to the right of 1 or to the left of O. The negative sign of

$$a_n x, \; a_n x + a_{n-1}, \; a_n x^2 + a_{n-1} x, \; \text{etc.,}$$

will signify that the direction of the lines is downward. Nor are any alterations necessary in order to include the case that several or all of the lines $a_0, a_1, \cdots a_n$ are directed downward and correspond to negative numbers. They are laid off on the y-axis in the same way as if to find the sum

$$a_0 + a_1 + a_2 + \cdots + a_n,$$

C_{a+1} lying above or below C_a according to a_a being directed upward or downward. The construction can be repeated for a number of values of x. The points P_0 will then represent the curve, whose equation is

$$y = a_0 + a_1 x + \cdots + a_n x^n,$$

x and y measuring abscissa and ordinates in independent units of length.

In order to draw the curve for large values of x a modification must be introduced. It will not do to choose $O1$ small in order to keep x on your drawing board; for then the lines $B_a C_a$ will become too short and thus their direction will be badly defined. The way to proceed is to change the variable. Write for instance $X = x/10$, so that X is ten times as small as x and write

$$A_a = a_a \cdot 10^a.$$

Then as

$$y = a_0 + a_1 \cdot 10 \frac{x}{10} + a_2 10^2 \frac{x^2}{10^2} + \cdots + a_n 10^n \frac{x}{10^n}$$

we find

$$y = A_0 + A_1 X + A_2 X^2 + \cdots + A_n X^n.$$

Lay off the lines A_0, A_1, $\cdots$ A_n in a convenient scale and let X play the part that x played before. The curve differs in scale from the first curve and the reduction of scale may be different for abscissas and ordinates but may if we choose be made the same so that it is geometrically similar to the first curve reduced to one tenth. It is evident that any other reduction can be effected in the same manner. By increasing the ratio x/X we enhance the value of A_n in comparison to the coefficients of lesser index, so that for the figure of the curve drawn in a very small scale all the terms will be insignificant except $A_n X^n$. In this case the points C_1, C_2, $\cdots$, C_n will very nearly coincide with O and only C_{n+1} will stand out.

It is interesting to observe that the best way of calculating an integral function

$$a_0 + a_1 x + a_2 x^2 + \cdots + a_n x^n$$

for any value of x proceeds on exactly the same lines as the geometrical construction. The coefficient a_n is first multiplied with x and a_{n-1} is added   Call the result a_{n-1}'. This is again multiplied by x and a_{n-2} is added. Call this result a_{n-2}'. Continuing in this way we finally obtain a value of a_0', which is equal to the value of the integral function for the value of x considered. Using a slide rule all the multiplications with x can be effected with a single setting of the instrument. The coefficients a_a and the values a_a' are best written in rows in this way

$$
\begin{array}{ccccc}
a_n & a_{n-1} & a_{n-2} & \cdots\; a_1 & a_0 \\
& a_n x & a_{n-1}'x & \cdots\; a_2'x & a_1'x \\
\hline
& a_{n-1}' & a_{n-2}' & \cdots\; a_1' & a_0'
\end{array}
$$

The accuracy of the slide rule is very nearly the same as the accuracy of a good drawing. But the rapidity is very much greater. When therefore only a few values of the integral function are required, the geometrical construction will not repay

the trouble.   It is different, however, when the object is to make
a drawing of the curve. The values supplied by calculation
would have to be plotted, while the geometrical construction
furnishes the points of the curve right away and in this manner
gains on the numerical method.

There is another geometrical method, which in some cases
may be just as good. Let us propose to find the value of an
integral function of the fourth degree.

$$y = a_0 + a_1x + a_2x^2 + a_3x^3 + a_4x^4$$

and let all coefficients in the first instance be positive.

The coefficients a_0, a_1, a_2, a_3, a_4 are supposed to be represented
by straight lines, while x will be the ratio of two lines.   The lines
a_0, a_1, a_2, a_3, a_4 are laid off in a
broken line a_0 to the right from
C_0 to C_1, a_1 upward from C_1 to
C_2, a_2 to the left from C_2 to C_3, a_3
downward from C_3 to C_4, a_4 again
to the right from C_4 to C_5 (Fig. 7).

Through C_5 draw a line C_5A to
a point A on C_3C_4 or its prolongation and let x be equal to the
ratio $C_4A : C_4C_5$ taken positive
when C_4A has the same direc-

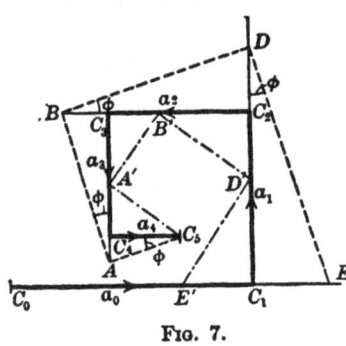

Fig. 7.

tion as C_3C_4.   Then we have

$$C_4A = a_4x,$$

and

$$C_3A = a_4x + a_3.$$

C_4A and C_3A are positive or negative according to their direction,
being the same as the direction of C_3C_4 or opposite to it.   Through
A draw the line AB forming a right angle with C_5A to a point B
on C_2C_3 or its prolongation.   Then we have

$$C_3B = C_3 A \cdot x = (a_4x + a_3) x$$

and

$$C_2B = a_4x^2 + a_3x + a_2.$$

C_3B and C_2B are positive or negative according to their direction being the same as the direction of C_2C_3 or opposite to it. Similarly we get

$$C_1D = a_4x^3 + a_3x^2 + a_2x + a_1,$$

and finally

$$C_0E = a_4x^4 + a_3x^3 + a_2x^2 + a_1x + a_0.$$

C_0E is positive, when E is on the right side of C_0 and negative when on the left side. When the point A moves along the line C_3C_4, the point E will move along the line C_0C_1 and its position will determine the values of the integral function. To find the position of E for any position of A, we might use transparent squared paper, that we pin onto the drawing at C_5, so that it can freely be turned round C_5. Following the lines of the squared paper

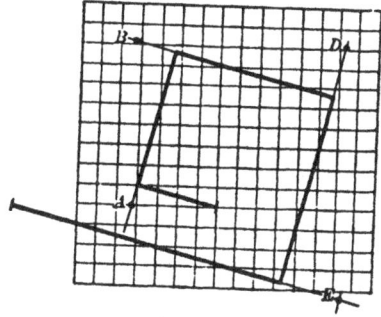

FIG. 8.

along C_5ABDE after turning it through a small angle furnishes the position of E for a new position of A (Fig. 8).

To include the case of negative coefficients we draw the corresponding line in the opposite direction. If for instance a_3 is negative C_3C_4 would have to lie above C_3; but C_3A would have to be counted in the same way as before, positive in a downward, negative in an upward direction.

The extension of the method to integral functions of any degree is obvious and need not be insisted on. It may be applied with advantage to find the real roots of an equation of any degree. For this purpose the broken line C_5ABDE would have to be drawn in such a way that E coincides with C_0. In the case of Fig. 7, for instance, it is easily seen that no real root exists. Fig. 9 shows the application to the quadratic equation. A circle is drawn over C_0C_3 as diameter. Its intersections with C_1C_2 furnish the points A and A' that correspond to the two roots. Both roots are negative in this case.

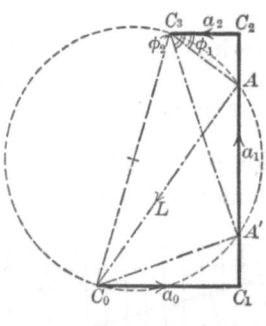

FIG. 9.

The first method of constructing the values of an integral function can be extended to the case where the function is given as the sum of a number of polynomials of the form

$$y = a_0 + a_1(x - p) + a_2(x-p)(x -q) \\ + a_3(x - p)(x - q)(x - r) + \cdots.$$

Let us again suppose $a_0, a_1, a_2, \cdots$ to represent straight lines laid off as before on the y-axis upwards or downwards as if to find their sum. $x, p, q, r \cdots$ are meant, to be numbers represented by the ratio of certain segments on the axis of abscissas. Let us consider the case of four terms, the highest polynomial being of the third degree. The fixed distance between the points marked p and $p + 1$, q and $q + 1$, r and $r + 1$ on the axis of abscissas, Fig. 10 is chosen arbitrarily and the position

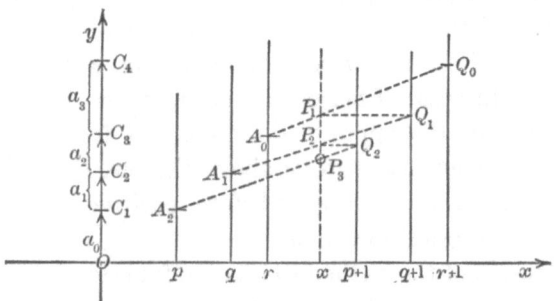

FIG. 10.

of the points marked p, q, r, x is made such that the ratio of Op, Oq, Or, Ox to that fixed distance is equal to the numbers p, q, r, x. For negative values the points are taken on the left of O.

Draw parallels to the y-axis through p, q, r, x, $p+1$, $q+1$, $r+1$. On the parallel through $r+1$ find the point Q_0 of the same ordinate as C_4 and on the parallel through r find the point A_0 of the same ordinate as C_3. Join A_0 and Q_0 by a straight line and find its intersection P_1 or that of its prolongation with the parallel through x. The height of P_1 above C_3 or A_0 is equal to $a_3(x-r)$ and the height above C_2 is equal to $a_3(x-r)$ $+ a_2$. On the parallel through $q+1$ find a point Q_1 of the same ordinate as P_1 and on the parallel through q a point A_1 of the same ordinate as C_2. Join A_1 and Q_1 by a straight line and find its intersection P_2 or that of its prolongation with the parallel through x. The height of P_2 above C_2 or A_1 is equal to

$$[a_3(x-r) + a_2](x-q),$$

and the height above C_1 is equal to

$$a_3(x-r)(x-q) + a_2(x-q) + a_1.$$

Finally find a point Q_2 on the parallel through $p+1$ of the same ordinate as P_2 and a point A_2 on the parallel through p of the same ordinate as C_1. Join A_2 and Q_2 by a straight line and find its intersection P_3 or that of its prolongation with the parallel through x. The height of P_3 above C_1 or A_2 will then be equal to

$$[a_3(x-r)(x-q) + a_2(x-q) + a_1](x-p)$$

and the ordinate of P_3 will be equal to the given integral function

$$y = a_3(x-r)(x-q)(x-p) + a_2(x-q)(x-p)$$
$$+ a_1(x-p) + a_0.$$

For large numbers p, q, r, x we use a similar device as before by introducing new numbers P, Q, R, X equal to one tenth, or one hundredth or any other fraction of $pqrx$. For instance

$$P = p/10, \quad Q = q/10, \quad R = r/10 \quad X = r/10.$$

We then write

$$A_0 = a_0, \quad A_1 = 10a_1, \quad A_2 = 100a_2, \quad A_3 = 1000a_3,$$

and obtain

$$y = A_0 + A_1(X - P) + A_2(X - P)(X - Q)$$
$$+ A_3(X - P)(X - Q)(X - R).$$

The scale for the lines A_0, A_1, A_2, A_3 and y must then be reduced conveniently and the values are constructed in the same way as before.

Now let us consider the inverse problem. The values of the integral function are given for

$$x = p, \; q, \; r, \; s;$$

find the lines a_0, a_1, a_2, a_3, so that the value of the integral function may be found for any other value of x in the way shown above.

Let us designate the given values of the integral function for $x = p, q, r, s$ by y_p, y_q, y_r, y_s and the points on the parallels through p, q, r, s with these ordinates by P, Q, R, S (see Fig. 12).

For $x = p$ the integral function

$$y = a_0 + a_1(x - p) + a_2(x - p)(x - q) + a_3(x - p)(x - q)(x - r)$$

reduces to a_0. Therefore we have $y_p = a_0$. The point C_1 is found by drawing a parallel to the axis of abscissas through P and taking its intersection with the axis of ordinates.

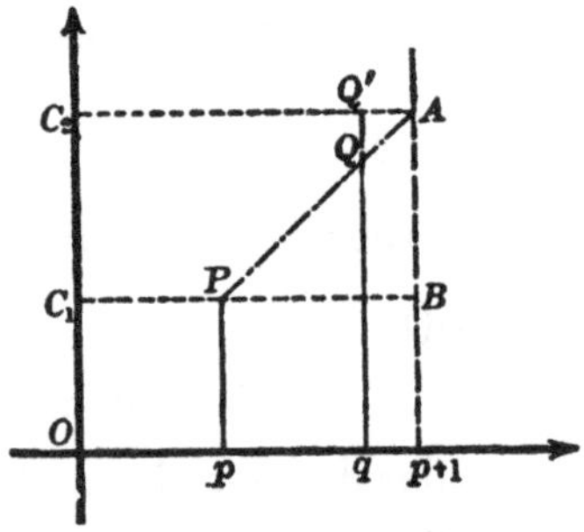

In order to find C_2 draw a straight line through P and Q and find its intersection A with the parallel through $p + 1$ (Fig. 11). A parallel to the axis of abscissas through A intersects the axis of ordinates in C_2. For the differences $y_q - y_p$ and $y_a - y_p$ (writing y_a for the ordinate of

FIG. 11.

A) are proportional to the differences of the abscissas and consequently in the ratio $(q - p) : 1$. Therefore

$$y_a - y_p = \frac{y_q - y_p}{q - p} = a_1.$$

In the same way as the point Q on the parallel through q we might join any point X on a parallel through x with the point P, find the intersection with the parallel through $p + 1$ and draw a parallel to the axis of abscissas. The point of intersection of

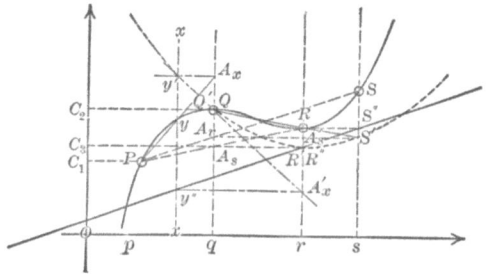

FIG. 12.

this parallel with the vertical through x let us call X' and its ordinate y'. Then we have

$$y' - y_p = \frac{y - y_p}{x - p} = a_1 + a_2(x - q) + a_3(x - q)(x - r).$$

Let us carry out this construction not only for $x = q$ but also for $x = r$ and $x = s$. This leads us to three points Q', R', S' on the verticals through q, r, s, whose ordinates are the values of the integral functions

$$y' = (a_0 + a_1) + a_2(x - q) + a_3(x - q)(x - r).$$

In this way we have reduced our problem. Instead of having to find an integral function of the third degree from four given points P, Q, R, S, we have now only to find an integral function of the second degree from three given points Q', R', S'. A second reduction is effected in exactly the same manner. Q' is joined with R' and S' by straight lines and through their intersection with the vertical through $q + 1$ parallels to the axis of abscissas are drawn that intersect the verticals through r and s in the points R'' and S'' respectively. The ordinates of these points are the values of the integral function y'' defined by

$$y'' - y_q' = \frac{y' - y_q'}{x - q} = a_2 + a_3(x - r),$$

for $x = r$ and $x = s$, or

$$y'' = a_0 + a_1 + a_2 + a_3(x - r).$$

The horizontal through R'' intersects the axis of ordinates in the point C_3. Finally we find C_4 by drawing a parallel to the axis of abscissas through the intersection of $R''S''$ or its prolongation with the vertical through $r + 1$.

Having found the points $C_1 C_2 C_3 C_4$ we can now for any value of x construct the ordinate

$$y = a_0 + a_1(x - p) + a_2(x - p)(x - q) \\ + a_3(x - p)(x - q)(x - r),$$

and thus draw the parabola of the third degree passing through the four points P, Q, R, S.

The construction may be somewhat simplified first by making $p + 1 = q$. Our data are the points P, Q, R, S, and we are perfectly at liberty to make the vertical through $p + 1$ coincide with the vertical through Q. In this case the point Q' will coincide with Q. · The parabola of the second degree through the points $Q'R'S'$ is again independent of the distance between the verticals through q and $q + 1$ and at the same time independent of the point P. Therefore we are perfectly at liberty, for the construction of any point of this parabola, to make the vertical through $q + 1$ coincide with the vertical through R even if the distance of the verticals through P and Q is different from that of the verticals through Q and R. R'' will in this case coincide with R'. The procedure is shown in Fig. 12. Starting from the points P, Q, R, S the first step is to find R', S' by connecting R and S with P and drawing horizontals through the intersections A_r, A_s with the vertical through q. The next step is to find S'' by connecting Q (identical with Q') with S' and drawing a horizontal through the intersection with the vertical through r. Now the straight line $R''S''$ can be drawn (R'' being identical with R'). On the vertical through any point x take the intersection with $R''S''$ and pass horizontally to the point A_x' on the vertical through r. Draw the line $Q'A_x'$ and find its intersection with the vertical through x. This point is on the parabola through $Q'R'S'$. Pass horizontally to the point A_x on the vertical through q and draw the line A_xP. Its intersection with the vertical through x is a point on the parabola of the third degree through P, Q, R, S.

The method is evidently applicable to any number of given points, the degree of the parabola being one unit less than the number of points.

The methods for the construction of the values of an integral function may be applied to find the value of any rational function

$$y = R(x).$$

For a rational function can always be reduced to the form of a quotient of two integral functions

$$R(x) = g_1(x)/g_2(x).$$

Now after having constructed curves whose ordinates give the values of $g_1(x)$ and $g_2(x)$ for any abscissa x (Fig. 13), $R(x)$ is found in the following manner.

Through a point P on the axis of abscissas draw a parallel to the axis of ordinates. Let G_1 and G_2 be the points whose ordinates are equal to $g_1(x)$ and $g_2(x)$. Pass horizontally from G_1 to G_1' on the vertical through P and

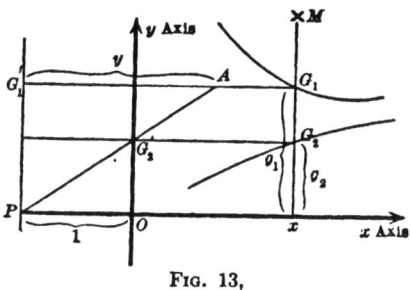

FIG. 13,

from G_2 to G_2' on the axis of ordinates. Draw a line through P and G_2' and produce it as far as A where it intersects the horizontal through G_1. Then $R(x)$ is equal to the ratio $G_1'A$ to PO. $G_1'A$ may then be set off as ordinate on the vertical

3

through x and defines the point M whose ordinate is equal to $R(x)$ in length, when OP is chosen as the unit of length.

§ 4. *Linear Functions of Any Number of Variables.*—Let us consider a linear function of a number of variables $x_1, x_2 \cdots x_n$,

$$a_0 + a_1x_1 + a_2x_2 + \cdots + a_nx_n,$$

where $a_0, a_1, a_2, \cdots a_n$ are given numbers positive or negative. The question is how the value of this linear function may be conveniently constructed for various systems $x_1, x_2, \cdots x_n$. Suppose $a_0, a_1, \cdots a_n$ to represent horizontal lines directed to the right or left according to the sign of the corresponding number and to be laid off on an horizontal axis in succession as if to find the sum

$$a_0 + a_1 + a_2 + \cdots + a_n,$$

a_0 begins at O and runs to C_1, a_2 begins at C_1 and runs to C_2 and so on (Fig. 14). The numbers $x_1, x_2, \cdots x_n$ let us represent

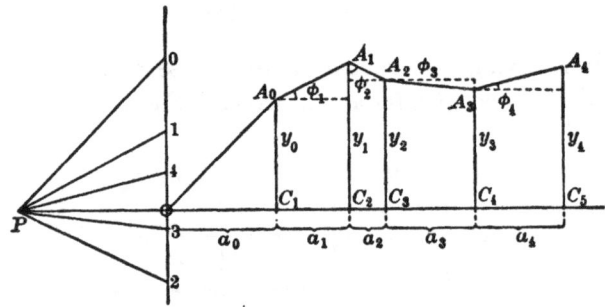

FIG. 14.

by ratios of lengths. We draw a vertical line through O and choose a point P on the horizontal axis. Then let x_1 be equal to the ratio $O1/PO$, $x_2 = O2/PO$, etc. If P is chosen on the left of O, we take the point 1 above O for a positive value of x_1 and below O for a negative one and the same for the other points. Mark a point 0 above O in the same distance from O as P. Join the point P with the points 0, 1, 2, 3, 4, $\cdots$ and draw a broken line $OA_0A_1A_2A_3A_4$ in such a manner that A_0 is on the vertical through C_1 and OA_0 is parallel to $P0$, A_1 on the vertical through C_2 and A_0A_1 parallel to $P1$, A_2 on the vertical through C_3 and A_1A_2 parallel to $P2$ and so on. Then the ordinate y_0 of A_0 will have the same length as a_0 and will be directed upward when the direction of a_0 is to the right, and downward when the direction of a_0 is to the left. The difference $y_1 - y_0$ of the ordinates of A_1 and A_0 is equal in length to a_1x_1, as $y_1 - y_0$ and a_1 have the same ratio as $O1$ and PO. A_1 will be above or below A_0 according to the line a_1x_1 being directed to the right or to the left and it is understood that a_1x_1 has the same direction as a_1 for positive

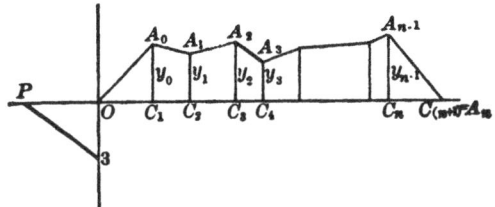

FIG. 15.

values of x_1 and a direction opposite to a_1 for negative values of x_1. Thus the ordinate y_1 has the same length as the line $a_0 + a_1x_1$ and its direction is upward or downward according to the direction of the line $a_0 + a_1x_1$ being to the right or to the left. In the same way it is shown that the ordinate y_2 of the point A_2 is equal in length to

$$a_0 + a_1x_1 + a_2x_2,$$

and y_3 to

$$a_0 + a_1x_1 + a_2x_2 + a_3x_3$$

and so on, the direction upward or downward corresponding to the positive or negative value of the linear function.

If the values of x_1, x_2, $\cdots$ x_n satisfy the equation

$$a_0 + a_1x_1 + a_2x_2 + \cdots + a_nx_n = 0$$

the ordinate y_n must vanish, that is to say, the point A_n must coincide with C_{n+1}, the end of the line a_n. And vice versa if A_n and C_{n+1} coincide the equation is satisfied. Consequently if we know all the values but one of the numbers x_1, x_2, $\cdots$ x_n the unknown value can be found graphically. For suppose x_3 to be

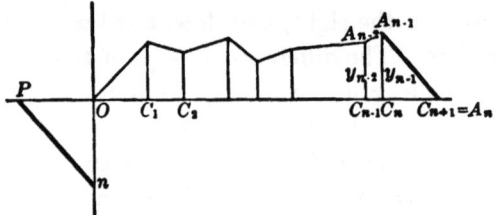

FIG. 16.

the unknown value we can, beginning from O, find the broken line as far as A_2 and beginning from the other end A_n we can find it as far as A_3 (Fig. 15). A parallel to A_2A_3 through P furnishes the point 3 on the axis of ordinates. If x_1, x_2, $\cdots$ x_{n-1} are known and only x_n not, we can draw the broken line as far as A_{n-1} and as A_n has to coincide with C_{n+1}, we can draw a parallel to $A_{n-1}A_n$ through P and find the point n on the axis of ordinates

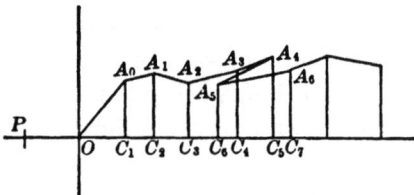

FIG. 17.

that determines the value x_n by the ratio On/PO or On/Oo. In Figs. 15 and 16 all the coefficients a_0, a_1, $\cdots$, are positive. A negative coefficient a_5 is shown in Fig. 17. The only difference is that C_6 lies to the left of C_5 and consequently the broken line passes from A_4 back to A_5.

If we keep the points 0, 1, 2, $\cdots$, in their positions but change the position of P to P' (Fig. 18) and repeat the construction of the broken line, we obtain $OA_0'A_1'A_2' \cdots$ instead of $OA_0A_1A_2 \cdots$. The ordinate y_a' of the point A_a' is evidently

$$y_a' = a_0 \frac{00}{P'0} + a_1 \frac{01}{P'0} + \cdots + a_a \frac{0a}{P'0}$$

and therefore

$$y_a' = \frac{P0}{P'0} y_a.$$

That is to say, by changing the position of P without changing the position of the points $0, 1, 2, \cdots$ we can change the scale of the ordinates of the broken line. They change inversely pro-

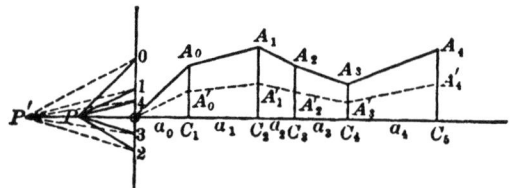

FIG. 18.

portional to PO. It may be convenient to make use of this device in order to make the ordinates a convenient size independent of the scale that we have chosen for the points $0, 1, 2, \cdots$ that determine the values

$$x_1 = \frac{01}{00}, \quad x_2 = \frac{02}{00}, \quad \cdots.$$

A linear equation with only one unknown quantity

$$a_0 + a_1x_1 = 0$$

is solved by drawing a parallel to A_0A_1 through P. Let a second equation be given with two unknown quantities

$$b_0 + b_1x_1 + b_2x_2 = 0.$$

The lines b_0, b_1, b_2 are laid off as before. Knowing x_1 as the solution of the first equation we can construct the broken line OB_0B_1 corresponding to the second equation and as B_2 must coincide with the end of b_2, we can draw a parallel to B_1B_2 through P and find x_2. In a similar manner we can find x_3 from a third equation

$$c_0 + c_1x_1 + c_2x_2 + c_3x_3 = 0,$$

and so we can find any number of unknown quantities, if each equation contains one unknown quantity more than those before.

In the general case when n unknown quantities are to be determined from n linear equations each equation will contain all the unknown quantities, and therefore we cannot find them one after the other as in the case just treated. But it can be shown that by means of very simple constructions the general case is reduced to a set of equations, such as has just been treated.

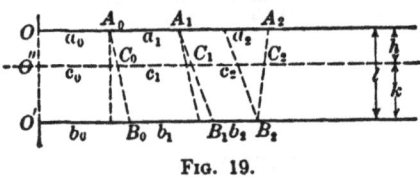

FIG. 19.

Let us begin with two equations and two unknown quantities.

$$a_0 + a_1x_1 + a_2x_2 = 0,$$
$$b_0 + b_1x_1 + b_2x_2 = 0.$$

The lines a_0, a_1, a_2 are laid off on a horizontal line $OA_0A_1A_2$ and the lines b_0, b_1, b_2 on another horizontal line $O'B_0B_1B_2$ (Fig. 19). Now let us join O and O', A_0 and B_0, A_1 and B_1, A_2 and B_2 by straight lines and let us draw a third horizontal line intersecting them in the points $O''C_0C_1C_2$. These points correspond to a certain linear function

$$c_0 + c_1x_1 + c_2x_2,$$

and it can be shown that it vanishes when x_1 and x_2 are the same values for which the first two linear functions vanish. Let the distance of the first two horizontal lines be l and the distance of the third from the first and second h and k. Then it can readily be seen that

$$c_0 = a_0 + \frac{h}{l}(b_0 - a_0) = \frac{k}{l}a_0 + \frac{h}{l}b_0.$$

For a parallel to OO' through A_0 defines with the line A_0B_0 on the third and second horizontal line segments equal to $c_0 - a_0$ and $b_0 - a_0$ and as these segments have the ratio h/l, it follows that

$$c_0 = a_0 + \frac{h}{l}(b_0 - a_0) = \frac{k}{l}a_0 + \frac{h}{l}b_0.$$

By drawing a parallel to A_0B_0 through A_1 and to A_1B_1 through A_2 or through B_2 (which comes to the same thing), we convince ourselves in the same way that

$$c_1 = a_1 + \frac{h}{l}(b_1 - a_1) = \frac{k}{l}a_1 + \frac{h}{l}b_2$$

and

$$c_2 = a_2 + \frac{h}{l}(b_2 - a_2) = \frac{k}{l}a_2 + \frac{h}{l}b_2.$$

Multiplying the equation

$$a_0 + a_1x_1 + a_2x_2 = 0$$

by k/l and the equation

$$b_0 + b_1x_1 + b_2x_2 = 0$$

by h/l and adding the two products, we obtain

$$c_0 + c_1x_1 + c_2x_2 = 0.$$

The third horizontal need not lie between the first two. If it lies below the second we have merely to give k a negative value and if it lies above the first we have to give h a negative value and the same formulæ for c_0, c_1, c_2 hold good. Consequently the conclusion remains valid, that from the first two equations the third follows.

Now as we are perfectly at liberty to draw the third horizontal line where we please, we can let it run through the intersection of the straight lines A_1B_1 and A_2B_2. In this case the points C_1 and C_2 must coincide and consequently c_2 must vanish. If c_1 does not vanish we can by what has been shown above find x_1 and with x_1 we can find x_2 from either of the two first horizontal lines. In case c_1 also vanishes, that is to say, in case the three straight lines A_2B_2, A_1B_1, A_0B_0 all pass through the same point, while OO' does not pass through it, the two given equations cannot simultaneously be satisfied. For if they were, it would follow that

$$c_0 + c_1x_1 + c_2x_2 = 0,$$

and as c_1 and c_2 are zero c_0 would have to be zero, which it is not as OO' is supposed not to pass through the intersection of A_2B_2, A_1B_1 and A_0B_0. If on the other hand all four lines A_2B_2, A_1B_1, A_0B_0, OO' pass through the same point, c_0, c_1 and c_2 will all three vanish. In this case the two given equations do not contradict one another, but $b_0b_1b_2$ will be proportional to $a_0a_1a_2$. The

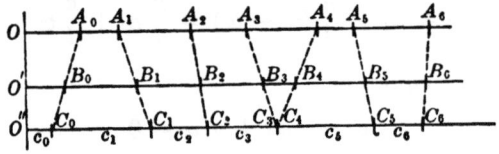

Fig. 20.

second equation will therefore contain the same relation between x_1 and x_2 as the first, so that there is only one condition for x_1 and x_2 to be satisfied. We may then assign any arbitrary value to one of them and determine the value of the other to satisfy the equation.

In the case of two linear equations of any number of quantities x_1, x_2, $\cdots x_n$ we can by the same graphical method eliminate one of the quantities. In Fig. 20 this is shown for two linear equations with six unknown quantities. The two horizontal lines $OA_0A_1A_2A_3A_4A_5A_6$ and $O'B_0B_1B_2B_3B_4B_5B_6$ represent two linear equations. Through the intersection of A_3B_3 and A_4B_4 a third horizontal line is drawn intersecting the lines OO', A_0B_0, A_1B_1, $\cdots A_6B_6$ in $O''C_0C_1 \cdots C_6$. As C_3 and C_4 coincide, the line c_4 vanishes and x_4 is eliminated, so that the equation assumes the form

$$c_0 + c_1x_1 + c_2x_2 + c_3x_3 + c_5x_5 + c_6x_6 = 0.$$

Suppose now that a set of six equations with six unknown quantities is represented geometrically on six horizontal lines. We shall keep one of these; but instead of the other five we construct five new ones from which one of the unknown quantities has been eliminated by means of the first equation. Now it may happen that at the same time another unknown quantity is eliminated, then this quantity remains arbitrary. Of the five new equations we again keep one that contains another unknown quantity and replace the four others again by four new ones from which this unknown quantity has been eliminated. Going on in this manner the general rule will be that with each step only one quantity is eliminated, so that at last one equation with one unknown quantity remains. Instead of the given six equations with six unknown quantities each, we now have one with six, one with five and so on down to one with one. The geometrical construction shows that this system is equivalent to the given system, for we can just as well pass back again to the given system. We have seen above how the unknown quantities may now be found geometrically. It may however happen in special cases that with the elimination of one unknown quantity another is eliminated at the same time. To this we may then assign an arbitrary value without interfering with the possibility of the solution. Finally all unknown quantities may be eliminated from an equation. If in this case there remains a term different from zero it shows that it is impossible to satisfy the given equations simultaneously. If no term remains, the two equations from which the elimination takes its origin contain the same relation between the unknown quantities and one of them may be ignored.

§ 5. *The Graphical Handling of Complex Numbers.*—A complex number

$$z = x + yi$$

is represented graphically by a point Z whose rectangular coördinates correspond to the numbers x and y. The units by which the coördinates are measured, we assume to be of equal length. We might also say that a complex number is nothing but an algebraical form of writing down the coördinates of a point in a plane. And the calculations with complex numbers stand for certain geometrical operations with the points which correspond to them.

By the "sum" of two complex numbers

$$z_1 = x_1 + y_1 i \quad \text{and} \quad z_2 = x_2 + y_2 i$$

we understand the complex number

$$z_3 = x_3 + y_3 i$$

where

$$x_3 = x_1 + x_2 \quad \text{and} \quad y_3 = y_1 + y_2,$$

and we write

$$z_3 = z_1 + z_2.$$

Graphically we obtain the point Z_3 representing z_3 from the points Z_1 and Z_2 representing z_1 and z_2 by drawing a parallel

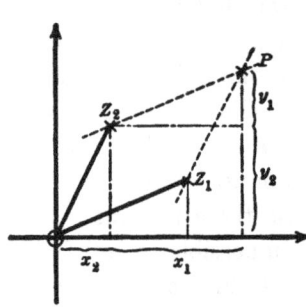

to OZ_2 through Z_1 and making Z_1P (Fig. 21) equal to OZ_2 in length *and direction* or by drawing a parallel through Z_2 and making Z_2P equal to OZ_1 in length and direction. The coordinates of P are evidently equal to $x_1 + x_2$ and $y_1 + y_2$.

Two complex numbers z and z' are called opposite, when their sum is zero.

FIG. 21.

$$z + z' = 0 \quad \text{or} \quad x = -x' \quad \text{and} \quad y = -y' \quad \text{or} \quad z = -z'.$$

The corresponding points Z and Z' are at the same distance from the origin O but in opposite directions.

The difference of two complex numbers is that complex number, which added to the subtrahend gives the minuend

$$z_2 + (z_1 - z_2) = z_1.$$

Therefore
$$z_1 - z_2 = (x_1 - x_2) + (y_1 - y_2)i.$$

This may also be written

$$z_1 + z_2' \quad \text{where} \quad z_2' = -z_2 = -x_2 - y_2i.$$

That is to say, the subtraction of the complex number z_2 from z_1 may be effected by adding the opposite number $-z_2$. For the geometrical construction of the point Z corresponding to $z_1 - z_2$ we have to draw a parallel to OZ_2 through Z_1 and from Z_1 in the direction from Z_2 to O we have to lay off the distance Z_2O. Or we may also draw from O a line equal in *direction* and in length to Z_2Z_1. This will also lead to the point Z representing the difference $z_1 - z_2$.

The rules for multiplication and division of complex numbers are best stated by introducing polar coördinates. Let r be the positive number measuring the distance OZ in the same unit of length in which x and y measure the abscissa and ordinate, so that

$$r = \sqrt{x^2 + y^2}$$

and let φ be the angle between OZ and the axis of x, counted in the direction from the positive axis of x toward the positive axis of y through the entire circumference (Fig. 22). Then we have

$$x = r \cos \varphi, \quad y = r \sin \varphi$$

and

$$z = x + yi = r(\cos \varphi + \sin \varphi i).$$

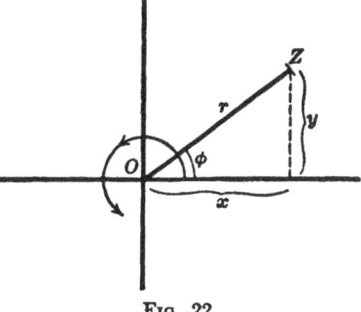

FIG. 22.

Let us call r the modulus and φ the angle of z. The angle may be increased or diminished by any multiple of four right angles without altering z, but any alteration of r necessarily implies an alteration of z.

According to Moivre's theorem, we can write

$$z = re^{\phi i}.$$

By the product of two complex numbers

$$z_1 = r_1 e^{\phi_1 i} \quad \text{and} \quad z_2 = r_2 e^{\phi_2 i}$$

we understand that complex number z_3 whose modulus r_3 is equal to the product of the moduli r_1 and r_2 and whose angle φ_3 is the sum of the angles φ_1 and φ_2 or differs from the sum only by a multiple of four right angles

$$z_3 = z_1 z_2 = r_1 r_2 e^{(\phi_1 + \phi_2)i}.$$

The definition of division follows from that of multiplication. The quotient z_1 divided by z_2 is that complex number, which multiplied by z_2 gives z_1. Therefore the product of its modulus with the modulus of z_2 must be equal to the modulus of z_1 and the sum of its angle with the angle of z_2 must be equal to the angle of z_1. Or we may also say the modulus of the quotient z_1/z_2 is equal to the quotient of the moduli r_1/r_2 and its angle is equal to the difference of the angles $\varphi_1 - \varphi_2$. An addition or subtraction of a multiple of four right angles we shall leave out of consideration as it does not affect the complex number nor the point representing it.

The geometrical construction corresponding to the multiplication and division of complex numbers is best described by considering two quotients each of two complex numbers that give the same result. Let us write

$$z_1/z_2 = z_3/z_4.$$

The geometrical meaning of this is that

$$r_1/r_2 = r_3/r_4,$$

and

$$\varphi_1 - \varphi_2 = \varphi_3 - \varphi_4.$$

That is to say, the triangles $Z_1 O Z_2$ and $Z_3 O Z_4$ are geometrically similar (Fig. 23). When three of the points Z_1, Z_2, Z_3, Z_4 are given the fourth can evidently be found. For instance let Z_1, Z_2, Z_4 be given. Draw a parallel to Z_1Z_2 intersecting OZ_2 at a distance r_4 from O. This point together with the intersection on OZ_1 and with O will form the three corners of a triangle congruent to the triangle Z_4Z_3O. It will be brought into

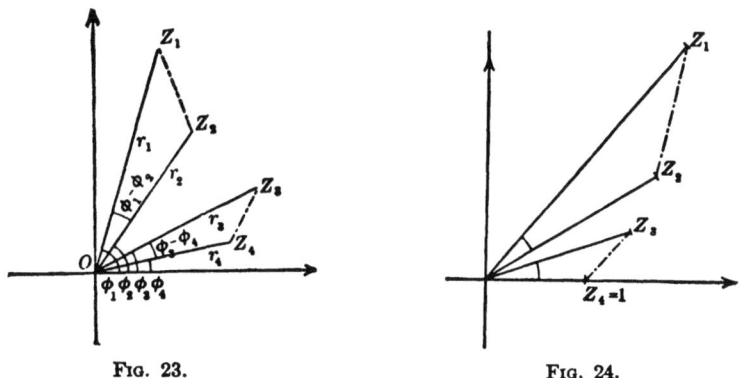

FIG. 23.   FIG. 24.

the position of Z_4Z_3O by being turned round O so as to bring the direction of the side in OZ_2 into the position of OZ_4. Thus the direction of OZ_3 and its length may be found.

This construction contains multiplication as well as division as special cases. Let Z_4 coincide with the point $x = 1$, $y = 0$, so that $z_4 = 1$ (Fig. 24), then we have

$$z_1/z_2 = z_3 \quad \text{or} \quad z_1 = z_2z_3.$$

From any two of the points Z_1, Z_2, Z_3 a simple construction gives us the third.

The geometrical representation of complex numbers may be used to advantage to show the properties of harmonic oscillations.

Let a point P move on the axis of x, so that its abscissa at the time t is given by the formula

$$x = r \cos (nt + \alpha),$$

n, r and α being constants. We call r the amplitude and $nt + \alpha$

the phase of the motion. The point P moves backwards and forwards between the limits $x = r$ and $x = -r$. The time $T = 2\pi/n$ is called the period of the oscillation, it is the time in which one complete oscillation backwards and forwards is performed.

Now instead of x let us consider the complex number

$$z = r \cos(nt + \alpha) + r \sin(nt + \alpha)i$$

or

$$z = re^{(nt+\alpha)i},$$

of which x is the abscissa and let us follow the movement of the point Z. For $t = 0$ we have

$$z = re^{\alpha i}.$$

Designating this value by z_0, we can write

$$z = z_0 e^{nti}.$$

The geometrical meaning of the product

$$z_0 e^{nti}$$

is that the line OZ_0 is turned round O through the angle nt. For the modulus of e^{nti} being equal to 1 the modulus of z_0 is not changed by the multiplication. The movement of the point Z therefore consists in a uniform revolution of OZ round O. At the moment $t = 0$ the position is OZ_0 and after the time $T = 2\pi/n$ the same position is occupied again. The revolution goes on in the direction from the positive axis of x to the positive axis of y (Fig. 25).

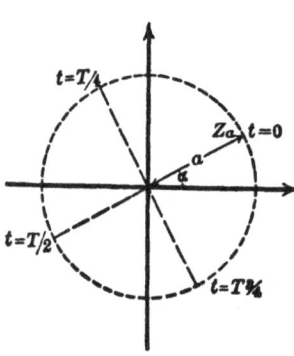

Fig. 25.

The movement of Z is evidently simpler than the movement of the projection P of Z on the axis of x.

Let us consider a motion composed of the sum of two harmonic motions of the same period but of different amplitudes and phases

$$x = r_1 \cos (nt + \alpha_1) + r_2 \cos (nt + \alpha_2),$$

and let us again substitute the motion of the point Z corresponding to the complex number

$$z = r_1 e^{(nt+\alpha_1)i} + r_2 e^{(nt+\alpha_2)i}.$$

For $t = 0$ the first term is

$$z_1 = r_1 e^{\alpha_1 i}$$

and the second term

$$z_2 = r_2 e^{\alpha_2 i}.$$

Introducing z_1 and z_2 into the expression for z we have

$$z = z_1 e^{nti} + z_2 e^{nti} = (z_1 + z_2)e^{nti} = z_3 e^{nti}$$

where

$$z_3 = z_1 + z_2.$$

This shows at once that the movement of Z is a uniform circular movement consisting in a uniform revolution of OZ round O. The position at the moment $t = 0$ is OZ_3 corresponding to the complex number

$$z_3 = z_1 + z_2.$$

The projection of Z on the axis of x has the abscissa

$$x = r_3 \cos (nt + \alpha_3)$$

where r_3 and α_3 designate modulus and angle of z_3. Thus the sum of two harmonic motions of the same period is shown also to form a harmonic motion.

The same holds for a sum of any number of harmonic motions of the same period. For the complex number

$$z = r_1 e^{(nt+\alpha_1)i} + r_2 e^{(nt+\alpha_2)i} + \cdots + r_\lambda e^{(nt+\alpha_\lambda)i}$$

where $r_1, r_2, \cdots r_\lambda$; $\alpha_1, \alpha_2, \cdots \alpha_\lambda$ and n are constants may be written

$$z = z_1 e^{nti} + z_2 e^{nti} + \cdots + z_\lambda e^{nti}$$

or

$$z = z_0 e^{nti},$$

where

$$z_0 = z_1 + z_2 + \cdots + z_\lambda.$$

The movement of Z therefore, excepting the case $z_0 = 0$, consists in a uniform revolution of OZ round O, OZ always keeping the same length equal to the modulus of z_0. The position of OZ at the moment $t = 0$ is OZ_0.

The motion of a point P whose abscissa is

$$x = ae^{-kt} \cos (nt + \alpha)$$

where a, k, n, α are constants (a and k positive) is called a damped harmonic motion. It may be looked upon as a harmonic motion, whose amplitude is decreasing. To study this motion let us again substitute a complex number

$$z = ae^{-kt} \cos (nt + \alpha) + ae^{-kt} \sin (nt + \alpha)i,$$

or

$$z = ae^{-kt} \cdot e^{(nt+\alpha)i},$$

or

$$z = z_0 e^{-kt} \cdot e^{nti},$$

where z_0 is written for the complex constant $ae^{\alpha i}$.

The product

$$z_0 e^{-kt}$$

is a complex number corresponding to a point Z_1 on the same radius as Z_0, coincident with Z_0 at the moment $t = 0$ but approaching O in a geometrical ratio after $t = 0$. In unit of time the distance of Z_1 from O decreases in the constant ratio $e^{-k} : 1$. The multiplication with e^{nti} turns OZ_1 round O through an angle nt. We may therefore describe the motion of Z as a uniform revolution of OZ round O, Z at the same time approaching O at a rate uniform in this sense that in equal times the distance is reduced in equal proportions (Fig. 26). At the moment $t = 0$ the position coincides with Z_0. We speak of a period of this motion meaning the time $T = 2\pi/n$ in which OZ performs an entire revolution round O, although it does not come back to its original position. Any part of the spiral curve described by Z in a given time is geometrically similar to any other part of the curve described in an interval of equal duration. For suppose the second interval of time happens τ units of time later, we shall have for the first interval

$$z = z_0 e^{-kt} \cdot e^{nti},$$

and for the second interval

$$z' = z_0 e^{-k(t+\tau)} \cdot e^{n(t+\tau)i}.$$

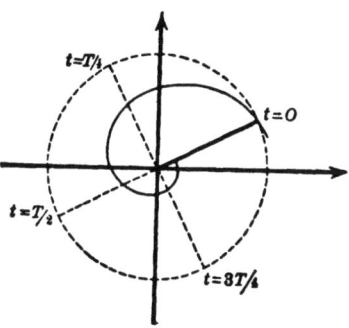

FIG. 26.

Now if z_1 and z_2 are the values of z at two moments t_1 and t_2 of the first interval and z_1' and z_2' the corresponding values of z' at the moments $t_1 + \tau$ and $t_2 + \tau$ of the second interval, we have

$$\frac{z_1}{z_2} = e^{-k(t_1-t_2)} \cdot e^{n(t_1-t_2)i} = \frac{z_1'}{z_2'}.$$

Therefore the triangle $Z_1 O Z_2$ is geometrically similar to the triangle $Z_1' O Z_2'$. As Z_1 and Z_2 may coincide with any points of the first part of the curve, the two parts are evidently geometrically similar.

The projection of Z on the axis of x performs oscillations decreasing in amplitude. The turning-points correspond to those points of the spiral curve described by Z, where its tangent is parallel to the axis of y, that is to say, where the abscissa of dz/dt vanishes.

Now

$$\frac{dz}{dt} = z_0(-k + ni)e^{-kt}e^{nti} = (-k + ni)z$$

or

$$\frac{\frac{dz}{dt}}{z} = -k + ni = \rho e^{i\lambda}$$

4

where ρ and λ are the modulus and angle of the complex number $-k + ni$.

Consequently, if we represent dz/dt by a point Z', the triangle $Z'OZ$ will remain geometrically similar to itself. The turning points of the damped oscillations correspond to the moments when OZ' is directed vertically upward or downward or when the angle of dz/dt is equal to $\pi/2$ or $3\pi/2$. The angle of z will then be $\pi/2 - \lambda$ or $3\pi/2 - \lambda$ plus or minus any multiple of 2π. As the angle of z, on the other hand, is changing in time according to the formula

$$nt + \alpha,$$

we find the moments where the movement turns by the equation

$$nt + \alpha = \pi/2 - \lambda + 2N\pi,$$

or

$$nt + \alpha = 3\pi/2 - \lambda + 2N\pi,$$

N denoting any positive or negative integral number. The time between two consecutive turnings is therefore equal to π/n, that is, equal to half a period. All the points Z corresponding to turning points lie on the same straight line through the origin O forming an angle $3\pi/2 - \lambda$ with the direction of the positive axis of x. The amplitudes of the consecutive oscillations therefore decrease in the same proportion as the modulus of z, that is to say, in half a period in the ratio $e^{-\frac{k\pi}{n}}$.

Let us consider the vibrations of a system possessing one degree of freedom when the system is subjected to a force varying as a harmonic function of the time and let us limit our considerations to positions in the immediate neighborhood of a position of stable equilibrium. If the quantity x determines the position of the system the oscillations satisfy a differential equation of the form

$$m \frac{d^2x}{dt^2} + k \frac{dx}{dt} + n^2 x = F \cos (pt)[1]$$

where m, k, n, p, F are positive constants.

[1] See for instance Rayleigh, Theory of Sound, Vol. I, chap. III, § 46.

This is another case where the introduction of a complex variable

$$z = x + yi$$

and the geometrical representation of complex numbers helps to form the solution and to survey the variety of phenomena that may be produced.

In order to introduce z let us simultaneously consider the differential equation

$$m\frac{d^2y}{dt^2} + k\frac{dy}{dt} + n^2y = F\sin pt,$$

and let us multiply the second equation by i and add it to the first. We then have

$$m\frac{d^2z}{dt^2} + k\frac{dz}{dt} + n^2z = Fe^{pti}.$$

The movement of the point Z representing the complex number z then serves as well to show the movement corresponding to x. We need only consider the projection of Z on the axis of x.

A solution of the differential equation may be obtained by writing

$$z = z_0 e^{pti}.$$

Introducing this expression for z and cancelling the factor e^{pti} we have

$$z_0(-mp^2 + kpi + n^2) = F,$$

or

$$z_0 = \frac{F}{-mp^2 + kpi + n^2}.$$

z_0 is a complex constant, that may be represented geometrically as we shall see later on.

This solution

$$z = z_0 e^{pti}$$

is not general. If z' denotes any other solution so that

$$m\frac{d^2z'}{dt^2} + k\frac{dz'}{dt} + n^2z' = Fe^{pti},$$

we find by subtracting the two equations

$$m\frac{d^2(z'-z)}{dt^2} + k\frac{d(z'-z)}{dt} + n^2(z'-z) = 0$$

or writing

$$z' - z = u,$$

$$m\frac{d^2u}{dt^2} + k\frac{du}{dt} + n^2u = 0.$$

The general solution of this equation is

$$u = u_1e^{\lambda_1 t} + u_2e^{\lambda_2 t},$$

where u_1 and u_2 are arbitrary constants and λ_1 and λ_2 are the roots of the equation for λ

$$m\lambda^2 + k\lambda + n^2 = 0,$$

$$\left.\begin{array}{c}\lambda_1\\\lambda_2\end{array}\right\} = -\frac{k}{2m} \pm \sqrt{\frac{k^2}{4m^2} - n^2}.$$

If $k^2/4m^2$ is greater than n^2, so that the square root has a real value, $\sqrt{k^2/4m^2 - n^2}$ will certainly be smaller than $k/2m$. Therefore λ_1 and λ_2 will both be negative and the moduli of the complex numbers $u_1e^{\lambda_1 t}$ and $u_2e^{\lambda_2 t}$ will in time become insignificant. If, on the other hand, $k^2/4m^2$ is smaller than n^2, both complex numbers $u_1e^{\lambda_1 t}$ and $u_2e^{\lambda_2 t}$ correspond to points describing spirals that approach the origin, as we have seen above, in a constant ratio for equal intervals of time. Therefore they will also in time become insignificant.

After a certain lapse of time the expression

$$z = z_0e^{pti}$$

will therefore suffice to represent the solution.

The point Z moves uniformly in a circle round O of a radius equal to the modulus of z_0, completing one revolution in the period $2\pi/p$, the period of the force acting on the system. The movement of the projection of Z on the axis of x is given by

$$x = r_0 \cos (pt + \alpha),$$

where r_0 is the modulus and α the angle of z_0. It is a harmonic movement with the same period as that of the force· $F \cos pt$, but with a certain difference of phase and a certain amplitude depending on the values of F, m, k, n, p.

It is important to study this relation in order to survey the phenomena that may be produced. For this purpose the geometrical representation of complex numbers readily lends itself.

In the expression for z_0

$$z_0 = \frac{F}{- mp^2 + kpi + n^2},$$

let us consider the denominator

$$- mp^2 + kpi + n^2,$$

and let us suppose the period of the force acting on the system not determined, while the constants of the system m, k, n and the amplitude of the force F have given values. The quantity p is the number of oscillations of the force during an interval of 2π units of time. This quantity p we suppose to be indeterminate and we intend to show how the amplitude and phase of the forced vibrations compare with the amplitude and phase of the force for different values of p.

Let us plot the curve of the points corresponding to the complex number

$$n^2 - mp^2 + kpi,$$

where p assumes the values $p = 0$ to $+ \infty$.

This curve is a parabola whose axis coincides with the axis of x and whose vertex is in the point $x = n^2$, $y = 0$. We find its equation by eliminating p from the equations

$$x = n^2 - mp^2, \quad y = kp,$$

viz.,

$$x = n^2 - \frac{m}{k^2} y^2.$$

But it is better not to eliminate p and to plot the different points for different values of p. In Fig. 27 the curve is drawn for $p = 0$

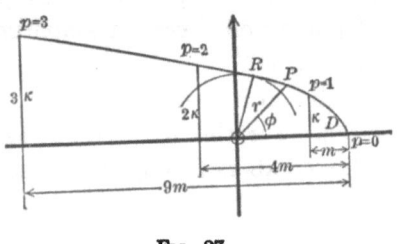

to 3 and the points for $p = 0, 1, 2, 3$ are marked. The ordinates increase in proportion to p; they are equal to $0, k, 2k, 3k$ for $p = 0, 1, 2, 3$. The distance between the projection of any point of the curve on the axis of x and

the vertex is proportional to p^2. It is equal to $0, m, 4m, 9m$ for $p = 0, 1, 2, 3$.

For any point P on the parabola let us denote the distance from O by r and the angle between OP and the positive axis of x by φ so that

$$n^2 - mp^2 + kpi = re^{\phi i}.$$

Then we have

$$z_0 = re^{\phi i},$$

and consequently

$$z = \frac{F}{r} e^{(pt-\phi)i},$$

and

$$x = \frac{F}{r} \cos (pt - \varphi).$$

The amplitude F/r of the forced vibration is inversely proportional to r. Thus our Fig. 27 shows us what the period of the force must be to make the forced vibrations as large as possible. It corresponds to the point on the parabola whose distance from O is smallest. It is the point where a circle round O touches the parabola. In Fig. 27 this point is marked R. It may be called the point of maximum resonance. When the constants of the system are such that the ordinate of the point, where the parabola intersects the axis of y is small in comparison with the abscissa of the vertex, then OR will lie close to the axis of y (Fig. 28). In this case the angle between OR and the positive axis of x will be very nearly equal to 90°, that is to say, the forced oscillations will lag behind the force oscil-lations by a little less than a quarter of a period. Keep-ing m and n constant, this will take place for small val-ues of k, *i. e.*, for a small damping influence. A small deviation of p from the fre-

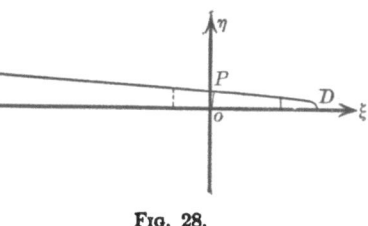

FIG. 28.

quency of maximum resonance will throw the point P away from R, so that r increases considerably and φ becomes either very small (for values of p smaller than the frequency of maximum resonance) or nearly equal to 180° (for values of p larger than the frequency of maximum resonance). In other words for small values of k the maximum of resonance is very sharp. A deviation of the period of the force from the period of maximum resonance will lessen the amplitude of the forced vibration considerably. The lag of its phase behind that of the force will at the same time nearly vanish, when the frequency of the force is decreased or it will become nearly as large as half a period, when the frequency of the force is in-creased. For larger values of k the parabola opens out and this phenomenon becomes less marked. The minimum of the radius r becomes less pronounced. The angle between OR and the axis of x becomes smaller and smaller and for a certain value of k and all larger values the point R will coincide with the vertex of the para-bola. In this case, there is no resonance. When the period of the force increases indefinitely (p becoming smaller and smaller) the amplitude of the forced vibration will increase and will approach more and more to the limit

$$\frac{F}{n^2},$$

but there will be no definite period for which the forced vibra-tions are stronger than for all others.

CHAPTER II.

The Graphical Representation of Functions of One or More Independent Variables.

§ 6. *Functions of One Independent Variable.*—A function y of one variable x

$$y = f(x)$$

is usually represented geometrically by a curve, in such a way that the rectangular coördinates of its points measured in certain chosen units of length are equal to x and y. This graphical representation of a function is exceedingly valuable. But there is another way not less valuable for certain purposes, more used in applied than in theoretical mathematics, which here will occupy our attention.

Suppose the values of y are calculated for certain equidistant values of x, for instance:

$$x = -6, -5, -4, -3, -2, -1, 0,$$
$$+1, +2, +3, +4, +5, +6,$$

and let us plot these values of y in a uniform scale on a straight line. Draw the uniform scale on one side of the straight line and mark the points that correspond to the calculated values of y on the other side of the straight line. Denote them by the numbers x that belong to them (Fig. 29). The drawing will then allow us to read off the value of y for any of the values of x with a certain accuracy depending on the size of the scale and the number of its partitions and naturally on the fine-

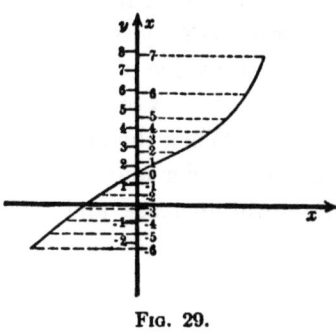

Fig. 29.

ness of the drawing. It will also allow us to read off the value of y for a value of x between those that have been marked, if the intervals between two consecutive values of x are so small that the corresponding intervals of y are nearly equal. We can with a certain accuracy interpolate values of x by sight. On the other hand, we can also read off the values of x for any of the values of y. We shall call this the representation of a function by a scale.

We can easily pass over to the representation of the same function by a curve. We need only draw lines perpendicular to the line carrying the scales through the points marked with the values of x and make their length measured in any given unit equal to the numbers x that correspond to them (Fig. 29).

In a similar way we can pass from the representation of the function by a curve to the representation by a scale.

The representation by a scale may be imagined to signify the movement of a point on a straight line, the values of x meaning the time and the points marked with these values being the positions of the moving point at the times marked. By passing over to the curve the movement in the straight line is drawn out into a curve with the time as abscissa (Fig. 30).

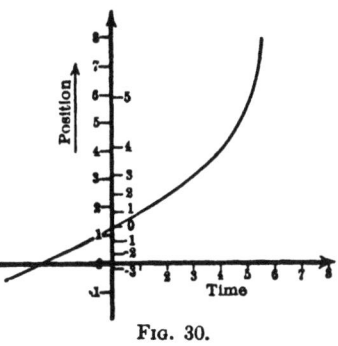

Fig. 30.

The representation by a scale is used in connection with the representation by a curve for the purpose of drawing a function of a function.

Let y be a function of x and x a function of t. Then we wish to represent y as a function of t.

Let $y = f(x)$ be given by a curve in the usual way and let $x = \varphi(t)$ be given by a scale on the axis of x marking the points where $t = 0, 1, 2, \cdots, 12$. We then find the values of y corresponding to the values $t = 0, 1, 2, \cdots, 12$ by drawing the ordinates of the curve $y = f(x)$ for the abscissas marked $t = 0, 1, 2,$

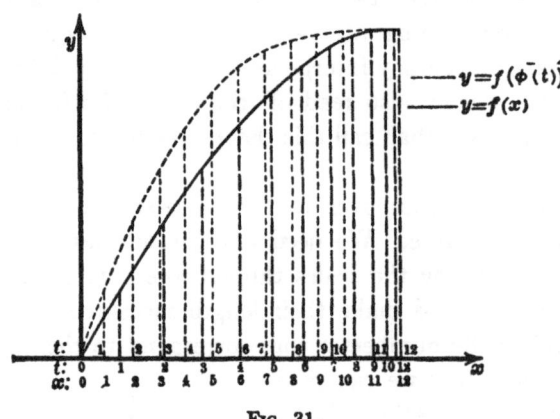

$\cdots, 12$. These ordinates as a rule will not be equidistant. But as soon as we move them so as to make them equidistant, they form the ordinates of the curve

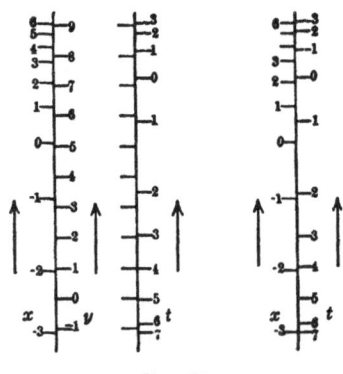

FIG. 32.

$$y = f(\varphi(t))$$

with t as abscissa (Fig. 31).

The representation of a function by a scale may be generalized in the respect that neither of the two scales facing one another on the straight line need necessarily be uniform. The intervals of both scales may vary from one side of the scale to the other. If the variation is sufficiently slow the interpolation can nevertheless be effected with accuracy. We may look at this case as composed of two cases of the first kind.

$$f(x) = y \quad \text{and} \quad y = g(t).$$

These scales are placed together, so that the scale x touches the scale t

$$f(x) = g(t),$$

while the scale y is cut out (Fig. 32).

§ 7. *The Principle of the Slide Rule.*—Let us investigate how the relation between x and t changes by sliding the x- and t-scales along one another.

If we slide the x-scale through an amount $y = c$ so that a point of the x-scale that was opposite to a certain point y of the y-scale, now is opposite $y + c$, then the relation between x and t represented by the new position of the scales will be given by the equation

$$f(x) = g(t) + c.$$

If x, t and x', t', denote two pairs of values that are placed opposite to one another, we shall have simultaneously

$$f(x) = g(t) + c,$$
$$f(x') = g(t') + c,$$

or by eliminating c

$$f(x) - g(t) = f(x') - g(t').$$

The ordinary slide rule carries two identical scales $y = \log x$ and $y = \log t$ that are able to slide along one another, x and t running through the values 1 to 100. We therefore have

$$\log x - \log t = \log x' - \log t',$$

or

$$\frac{x}{t} = \frac{x'}{t'}.$$

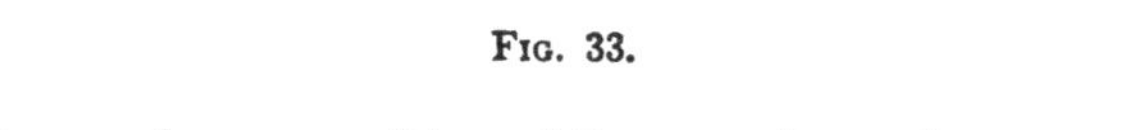

Fig. 33.

That is to say, in any position of the x- and t-scale any two values x and t opposite each other have the same ratio (Fig. 33). This

is the principle on which the use of the slide rule is founded.
It enables us to calculate any of the four quantities x, t, x', t'
if the other three are given. Suppose, for example, x, t, x'
known. We set the scales so that x appears opposite to t,

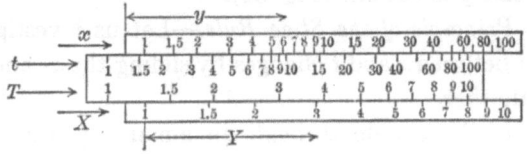

Fɪɢ. 34.

then t' is read off opposite to x'. On the other edges the slide
rule carries two similar scales one double the size of the other
(Fig. 34). We may write

$$y = 2 \log X \quad \text{and} \quad y = 2 \log T.$$

By means of a little frame carrying a crossline and sliding over
the instrument, we can bring the scales x and T or t and X opposite each other. If, for example, for any position of the
instrument x, T and x', T' are two pairs of values opposite each
other, then

$$\log x - 2 \log T = \log x' - 2 \log T',$$

or

$$\frac{x}{T^2} = \frac{x'}{T'^2}.$$

If any three of the four quantities x, T, x', T' are known the
fourth may be read off. Thus we find the value

$$\frac{x T'^2}{T^2},$$

by setting T opposite to x and reading off the value opposite to
T'. Or we can find the value of

$$\sqrt{\frac{x'}{x}}\, T$$

by setting x opposite to T and reading off the value opposite x'.

Let us reverse the part that carries the scales t, T so that x slides along T and X along t, but in the opposite order (Fig. 35).

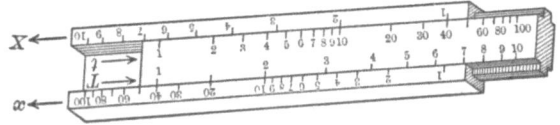

FIG. 35.

The scales t, T may then be expressed by

$$y = l - \log t \quad \text{and} \quad y = l - 2 \log T,$$

l being the entire length of the scales.

By setting the instrument to any position and considering the scales x and t or X and T by means of the cross line we have
$$\log x + \log t = \log x' + \log t' \quad \text{and} \quad \log X + \log T = \log X' + \log T'$$
or
$$xt = x't' \quad \text{and} \quad XT = X'T',$$

so that any two values opposite to one another have the same product.

For x and T we have

$$\log x + 2 \log T = \log x' + 2 \log T',$$
or
$$xT^2 = x'T'^2.$$

Let us apply this to find the root of an equation of the form

$$u^3 + au = b.$$

Divide by u so that

$$u^2 + a = \frac{b}{u}.$$

and set $T = 1$ opposite to $X = b$. Then taking $T = u$ we find on the same cross line $t = u^2$ and $X = b/u$, so that we read the two values u^2 and b/u directly opposite to each other on the scales t and X. If b/u is positive, it decreases while u^2 increases.

Running our eye along we have to find the place where the difference $b/u - u^2$ is equal to a. Having found it the T-scale gives us the root of the equation. For example take

$$u^3 - 5u = 3,$$

or

$$u^2 - 5 = \frac{3}{u}.$$

We set $T = 1$ opposite $X = 3$ and run our eye along the scales X and t (Fig. 36), to find the place where $t - 5 = X$. We find

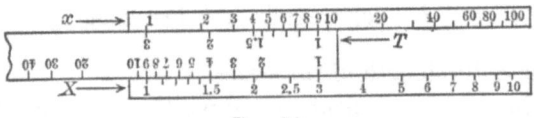

Fig. 36.

it approximately at $t = 6.2$, and on the T-scale we read off $T = 2.50$ as the approximate value of the root. This is the only positive root. But for a negative root $3/u$ is negative, and therefore the positive value of $3/u$ plus u^2 would have to be equal to 5. We run our eye along and find $t = .3.37$ opposite to $X = 1.63$, approximately corresponding to $T = 1.84$. Therefore $- 1.84$ is another root. As the coefficient of u^2 in the first form of the equations vanishes it follows that the sum of the three roots must be equal to zero. This demands a second negative root approximately equal to $- 0.66$. To make sure that it is so, we set the instrument back and take the other end of the T-scale as representing the value $T = 1$ and give it the position this end had before. Running our eye along the scales X and t, we find $t = 0.43$ opposite to $X = 4.57$, giving $X + t = 5.00$. On the T-scale we find 0.655, so that the third root is found equal to $- 0.655$.

When b is negative there is always one and only one negative root. For u running through the values $u = 0$ to $- \infty$, $u^2 - b/u$ will run from $- \infty$ to $+ \infty$ without turning. When b is positive there is always one and only one positive root; for then $u^2 - b/u$

runs from $-\infty$ to $+\infty$ for $u = 0$ to $+\infty$. In the first case there may be two positive roots or none; in the second case there may be two negative roots or none. For positive values of a one root only exists in either case. This is easily seen in the first form of the equation

$$u^3 + au = b,$$

because from a positive value of a it follows that $u^3 + au$ will for $u = -\infty$ to $+\infty$, run from $-\infty$ to $+\infty$ without turning and will therefore pass any given value once only.

In order to decide whether in the case of a negative value of a there are three roots or only one let us write

$$u^2 - \frac{b}{u} = -a.$$

For negative values of b we have to investigate whether there are positive roots. For positive values of u the function $u^2 - b/u$ has a minimum, when the differential coefficient vanishes, *i. e.,* for

$$2u + \frac{b}{u^2} = 0,$$

or

$$2u^2 = -\frac{b}{u}.$$

Having set our slide rule so that t gives us u^2 and X gives us $-b/u$, we find the value u where the minimum takes place by running our eye along and looking for the values X, t opposite each other for which X is twice the value of t

$$2t = X.$$

Then $t + X$ is the minimum of $u^2 - b/u$, so that there will be two or no positive roots according to $t + X$ being smaller or larger than $-a$. For positive values of b, we have to find out whether there are negative roots. The criterion is the same. After having set $T = 1$ opposite to b and having found the positive root, we find the place where

$$2t = X.$$

Then $t + X$ is the minimum of all values that $u^2 - b/u$ assumes for negative values of u. If the minimum is smaller than $- a$ there are two negative roots; if it is larger there are none. If it is equal to $- a$ the two negative roots coincide.

For the equation

$$u^2 - 5 = \frac{3}{u},$$

for instance, we find $t = 1.31$ opposite to $X = 2.62$ (Fig. 36), so that $2t = 2.62 = X$. Now $t + X = 3.93$ is smaller than 5, therefore $u^2 - 3/u$ will assume the value 5 for two negative values of u on either side of the value $u = - T = - 1.143$ for which the minimum of $u^2 - 3/u$ takes place.

On the same principle as the slide rule many other instruments may be constructed for various calculations. In all these cases we have for any position of the instrument

$$f(x) - g(t) = f(x') - g(t'),$$

where x, t are any readings of the two scales opposite each other and $x't'$ the readings at any other place. $f(x)$ and $g(t)$ may be any functions of x and t. It will only be desirable that they be limited to intervals of x and t, which contain no turning points. Else the same point of the scale corresponds to more than one value of x or t and that will prevent a rapid reading of the instrument.

Let us design an instrument for the calculation of the increase of capital at compound interest at a percentage from 2 per cent. upward. If x is the number of per cent. and t the number of years, the increase of capital at compound interest is in the proportion

$$\left(1 + \frac{x}{100}\right)^t.$$

We can evidently build an instrument for which

$$\left(1+\frac{x}{100}\right)^{t} = \left(1+\frac{x'}{100}\right)^{t'}.$$

For taking first the logarithm and then the logarithm of the logarithm, we obtain

$$\log t + \log \log \left(1+\frac{x}{100}\right) = \log t' + \log \log \left(1+\frac{x'}{100}\right).$$

We have only to make the x-scale

$$y = + \log \log \left(1+\frac{x}{100}\right) - \log \log \left(1+\frac{2}{100}\right),$$

and the t-scale

$$y = \log n - \log t.$$

For $x = 2$ we have $y = 0$ and therefore in the normal position of the instrument $t = n$. On the other end we have $t = 1$ and therefore $y = \log n$. Now let us take $n = 100$, so that $y = 2$ for $t = 1$. Say the length of the instrument is to be about 24 cm., then the unit of length for the y-scale would have to be 12 cm. In the normal position of the instrument the readings x, t opposite to each other satisfy the equation

$$\left(1+\frac{x}{100}\right)^{t} = \left(1+\frac{2}{100}\right)^{100}.$$

Opposite $t = 1$, we read the value $x_1 = 624$ and this gives us

$$\left(1+\frac{2}{100}\right)^{100} = 1+\frac{x_1}{100} = 1 + 6.24 = 7.24.$$

A capital will increase in 100 years at two per cent. compound interest in the proportion $7.24 : 1$. Or we may also say the number $x_1 = 624$ read off opposite $t = 1$ is the amount which is added to a capital equal to 100 by double interest of 2 per cent. in 100 years. The same position of the instrument gives us the number of years that are wanted for the same increase of capital

5

at a higher percentage.   For all the values x, t opposite to each other satisfy the equation

$$\left(1 + \frac{x}{100}\right)^t = 7.24.$$

For any other given percentage x and any other given number of years t the increase of capital is found by setting x opposite to t and reading the x-scale opposite to $t = 1$.   The only restriction is that the ratio is not greater than 7.24, else $t = 1$ will lie beyond the end of the x-scale.

For a given increase of capital the instrument will enable us either to find the number of years if the percentage is given, or the percentage if the number of years is given, subject only to the restriction mentioned.

We can build our instrument so as to include greater increases of capital by choosing a larger value of n. $n = 1000$, for instance, will make $y = 3$ for $t = 1$.   If the instrument is not to be increased in size the scales would have to be reduced in the proportion 2 : 3.

Let us consider another instance

$$y = \frac{1}{x}, \quad y = \frac{1}{n} - \frac{1}{t}.$$

In the normal position of the instrument the scale division marked $x = \infty$ corresponds to $y = 0$ and is opposite to $t = n$. If we have $t = \infty$ on the other end, the length of the instrument will correspond to $y = 1/n$.   Let us choose $n = 0.1$, so that the length of the instrument is $y = 10$.   That is to say, the unit of length of the y-scale is one tenth of the length of the instrument. For any position of the instrument we have

$$\frac{1}{x} + \frac{1}{t} = \frac{1}{x'} + \frac{1}{t'}.$$

If the scale division marked $x = \infty$ is opposite to $t = c$ we can write $x' = \infty$, $t' = c$ and have

$$\frac{1}{x} + \frac{1}{t} = \frac{1}{c}.$$

The instrument will therefore enable us to read off any one of the three quantities x, t, c, if the other two are given, the only restriction being that all three lie within the limits 0.1 to ∞. The instrument may be used to determine the combined resistance of two parallel electrical re-sistances, for the resistances satisfy the equation

$$\frac{1}{R} = \frac{1}{R_1} + \frac{1}{R_2}.$$

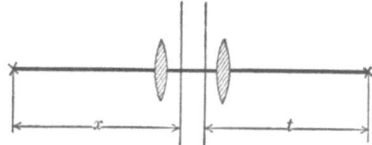

FIG. 37.

Similarly it may be used to calculate the distances of an object and its image from the principal planes of any given system of lenses. For if f is the focal length and x and t the distances of the object and its image from the corresponding principal planes (Fig. 37), the equation is

$$\frac{1}{x} + \frac{1}{t} = \frac{1}{f}.$$

On the back side of the movable part of an ordinary slide rule there generally is a scale

$$y = 2 + \log \sin t.$$

When this part is turned round and the scale is brought into contact with the scale

$$y = \log x,$$

we obtain for any position of the instrument

$$\log x - \log \sin t = \log x' - \log \sin t',$$

or

$$\frac{x}{\sin t} = \frac{x'}{\sin t'},$$

for any two pairs of values x, t that are opposite each other.

Given two sides of a triangle and the angle opposite the larger of the two the instrument gives at once the angle opposite the other side. Similarly when two angles and one side are given, it gives the length of the other side.

If $x' = a$ is the value opposite to $t' = 90°$, we have

$$x = a \sin t.$$

Thus we can read the position of any harmonic motion for any value of the phase.

An instrument carrying the scales

$$y = \log \sin x \quad \text{and} \quad y = \log \sin t$$

enables us to find any one of four angles x, t, x', t' for which

$$\frac{\sin x}{\sin t} = \frac{\sin x'}{\sin t'}$$

if the other three are given. Thus, knowing the declination, hour angle and height of a celestial body, we can read the azimuth on the instrument. We have only to take $x = 90°$ — height, t = hour angle, $x' = 90°$ — declination, then t' = azimuth or $180°$ — azimuth.

It is not necessary to carry out the subtraction $90°$ — height and $90°$ — declination. The difference may be counted on the scale by imagining $0°$ written in the place of $90°$, $10°$ in the place of $80°$ and so on and counting the partitions of the scale backwards instead of forward.

§ 8. *Rectangular Coördinates with Intervals of Varying Size.*— The two methods of representing the relation between two variables either by a curve connecting the coördinates or by scales facing each other lead to a combination of both.

Suppose the rectangular coördinates x and y are functions of u and v,

$$x = \varphi(u) \quad \text{and} \quad y = \psi(v).$$

The function $x = \varphi(u)$ is represented by a uniform scale for x on the axis of abscissæ facing a non-uniform scale for u. The function $y = \psi(v)$ is represented by a uniform scale for y on the axis of ordinates facing a non-uniform scale for v. Through the scale-divisions u let us draw vertical lines, and through the scale-divisions v let us draw horizontal lines. These two systems of parallel lines form a network of rectangular meshes of various sizes (Fig. 38), and any equation between u and v may be represented by a curve in this plane.

The usefulness of this method will be seen by some examples. It enables us by a clever choice of the functions $\varphi(u)$ and $\psi(v)$

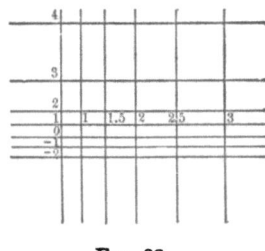

Fig. 38.

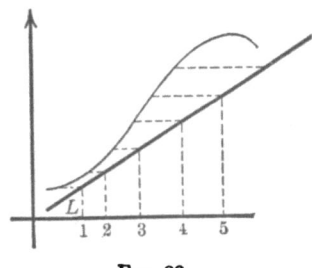

Fig. 39.

to simplify the form of the curve. It is easily seen, for instance, that a curve representing an equation $f(u, v) = 0$ may always be replaced by a straight line, if we choose the u-scale properly. For when the points $u = 1, 2, 3, 4, \cdots$ of the curve are not on a straight line, let them be moved to a straight line without altering their ordinates (Fig. 39). This will change the u-scale but it will not alter the equation $f(u, v) = 0$ now represented by the straight line.

Suppose we want to represent the relation

$$\frac{u^2}{a^2} + \frac{v^2}{b^2} = 1,$$

where a and b are given numbers. If u and v were ordinary rectangular coördinates the curve would be an ellipse. But if we make

$$x = u^2 \quad \text{and} \quad y = v^2$$

the equation of the line in rectangular coördinates becomes

$$\frac{x}{a^2} + \frac{y}{b^2} = 1,$$

and the curve will therefore be a straight line running from a point on the positive axis of x to a point on the positive axis of y. The point on the axis of x corresponds to the value $u = \pm a$ on the u-scale, and the point on the axis of y corresponds to the value $v = \pm b$ on the v-scale (Fig. 40).

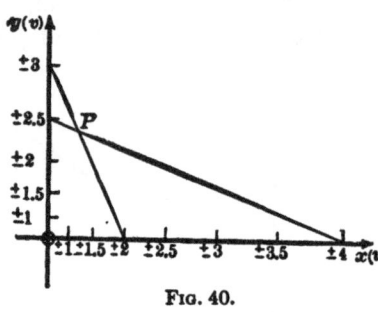

Fig. 40.

Any point on the straight line corresponds to four combinations $+u, +v; -u, +v; u, -v; -u, -v$, because x has the same values for opposite values of u and y for opposite values of v. We can read v as a function of u or u as a function of v.

If a second equation

$$\frac{u^2}{a_1^2} + \frac{v^2}{b_1^2} = 1$$

is given, we find the common solutions of the two equations by the intersection of the corresponding straight lines. Fig. 40 shows the solutions of the two equations

$$\frac{u^2}{2^2} + \frac{v^2}{3^2} = 1$$

and

$$\frac{u^2}{4^2} + \frac{2^2 v^2}{5^2} = 1,$$

approximately equal to $u = \pm 1.2$ and $v = \pm 2.4$.

Another function much used in mathematical physics

$$v = a e^{-\frac{u^2}{m^2}}$$

may also be represented by a straight line by means of the same device.

By making

$$y = \log v, \quad x = u^2,$$

we obtain

$$y = \log a - \frac{x}{m^2},$$

where $\log v$ and $\log a$ are the natural logarithms of v and a. The u-scale is laid off on the axis of x and the v-scale on the axis of y and we have to join the points $u = 0$, $v = a$ and $u = m$, $v = a/e$. The point $v = a/e$ is found by laying off the distance $v = 1$ to $v = e$ from $v = a$ downward (Fig. 41). We are not obliged to take the same units of length for x and y.

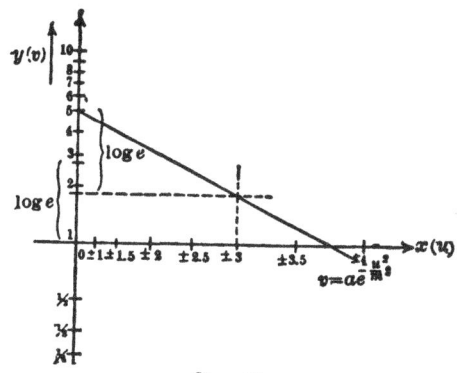

Fig. 41.

Suppose we had to find the constants a and m from two equations

$$v_1 = ae^{-\frac{u_1^2}{m^2}}$$

and

$$v_2 = ae^{-\frac{u_2^2}{m^2}}.$$

Our diagram would furnish two points corresponding to u_1, v_1 and u_2, v_2. The straight line joining these two points intersects the axis of ordinates at $v = a$ and intersects the parallel through $v = a/e$ to the axis of abscissæ at $u = m$.

In applied mathematics the problem would as a rule present itself in such a form that more than two pairs of values u, v would be given but all of them affected with errors of observation. The way to proceed would then be to plot the corresponding points and to draw a straight line through the points as best we can. A black thread stretched over the drawing may be used to advantage to find a straight line passing as close to the points as possible (Fig. 42).

In several other cases the variables u and v are connected with the rectangular coördinates x and y by the functions

$$x = \log u \quad \text{and} \quad y = \log v.$$

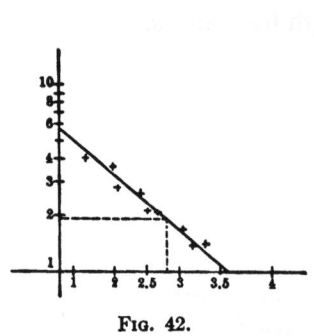

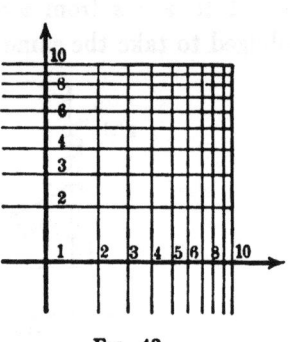

FIG. 42.　　　　　　　　　　FIG. 43.

"Logarithmic paper" prepared with parallel lines for equidistant values of u and lines perpendicular to these for equidistant values of v is manufactured commercially (Fig. 43).

By this device diagrams representing the relation

$$u^r v^s = c,$$

where r, s, c are constants are given by straight lines. For by taking the logarithm we obtain

$$rx + sy = \log c.$$

The straight line connects the point $u = c^{1/r}$ on the u-scale with the point $v = c^{1/s}$ on the v-scale.

Logarithmic paper is further used to advantage in all those

cases where a variety of relations between the variables u and v are considered that differ only in u and v being changed in some constant proportion. If u and v were plotted as rectangular coördinates the curves representing the different relations between u and v might all be generated from one of them by altering the scale of the abscissæ and independently the scale of the ordinates, so that the appearance of all these curves would be very different. Let us write

$$f(u, v) = 0,$$

as the equation of one of the curves. The equations of all the rest may then be written

$$f\left(\frac{u}{a}, \frac{v}{b}\right) = 0,$$

where a, b are any positive constants. The points u, v of the first curve lead to the points on one of the other curves by taking u a times as great and v b times as great. For if we write $u' = au$ and $v' = bu$ the equation $f(u, v) = 0$ leads to the equation between u' and v':

$$f\left(\frac{u'}{a}, \frac{v'}{b}\right) = 0.$$

Using logarithmic paper the diagram of all these curves becomes very much simpler. The equation $f(u, v) = 0$ is equivalent to a certain equation $\varphi(x, y) = 0$, where $x = \log u$, $y = \log v$. Now let x', y' be the rectangular coördinates corresponding to u', v' so that

$$x' = \log u' = \log u + \log a = x + \log a,$$

$$y' = \log v' = \log v + \log b = y + \log b.$$

The point x', y' is reached from the point x, y by advancing through a fixed distance $\log a$ in the direction of the axis of x and a fixed distance $\log b$ in the direction of the axis of y. The whole curve

$$f(u, v) = 0$$

drawn on logarithmic paper is therefore identical with all the curves

$$f\left(\frac{u}{a}, \frac{v}{b}\right) = 0.$$

It can be made to coincide with any one of the curves by moving it along the directions of x and y.

§ 9. *Functions of Two Independent Variables.*—When a function of one variable $y = f(x)$ is represented by a curve, the values of x are laid off on the axis of x and the values of y are represented by lines perpendicular to the axis of x. In a similar way a function of two independent variables

$$z = f(x, y)$$

may be represented by plotting x and y as rectangular coördinates and erecting lines perpendicular to the xy plane, in all the points x, y, where $f(x, y)$ is defined and making the lengths of the perpendiculars proportional to z. In this way the function corresponds to a surface in space. Now there are practical difficulties in working with surfaces in space and therefore it appears desirable to use other methods, that enable us to represent functions of two independent variables on a plane. This may be done in the following way.

Taking x, y as rectangular coördinates all the points for which $f(x, y)$ has the same value form a curve in the xy plane. Let us suppose a number of these curves drawn and marked with the value of $f(x, y)$. If the different values of $f(x, y)$ are chosen sufficiently close, so that the curves lie sufficiently close in the part of the xy plane that our drawing comprises, we are not only able to state the value of $f(x, y)$ at any point on one of the drawn curves, but we are also able to interpolate with a certain degree of accuracy the value of $f(x, y)$ at a point between two of the curves. As a rule it will be convenient to choose equidistant values of $f(x, y)$ to facilitate the interpolation of the values between. The curves may be regarded as the perpendicular projection of certain curves on the surface in space, the intersections of the surface by equidistant planes parallel to the xy plane.

The method is the generalization of the scale-representation of a function of one variable. For a relation between t and x represented by a curve with t as ordinate and x as abscissa, is transformed into a scale representation by perpendicularly projecting certain points of the curve onto the axis of x, the intersections of the curve by equidistant lines parallel to the axis of x and marking them with the value of t. A scale division in the case of a function of one variable corresponds to a curve in the case of a function of two independent variables.

This method of representing a function of two independent variables by a plane drawing or we might also say of representing a surface in space by a plane drawing, is used by naval architects to render the form of a ship and by surveyors to render the form of the earth's surface and by engineers generally. Let us apply the method to a problem of pure mathematics.

The equation

$$z^3 + pz + q = 0$$

defines z as a function of p and q. Let us represent this function by taking p and q as rectangular coördinates and drawing the lines for equidistant values of z.

For any constant value of z we have a linear equation between the variables p and q, and therefore it is represented by a straight line. This line intersects the parallels $p = 1$ and $p = -1$ at the points $q = -z^3 - z$ and $q = -z^3 + z$. Let us calculate these values for $z = 0$; ± 0.1; $\pm 0.2 \cdots \pm 1.3$ and in this way draw the lines corresponding to these values of z as far as they lie in a square comprising the values $p = -1$ to $+1$ and $q = -1$ to $+1$. Fig. 44 shows the result. On this diagram we can at once read the roots of any equation of the third degree of the form

$$z^3 + pz + q = 0,$$

where p and q lie within the limits -1 to $+1$. For $p = 0.4$ and

$q = -0.2$, for instance, we read $z = 0.37$, interpolating the value of z according to the position of the point between the lines $z = 0.3$ and $z = 0.4$. We also see that there is only one real root, for there is only one straight line passing through the point.

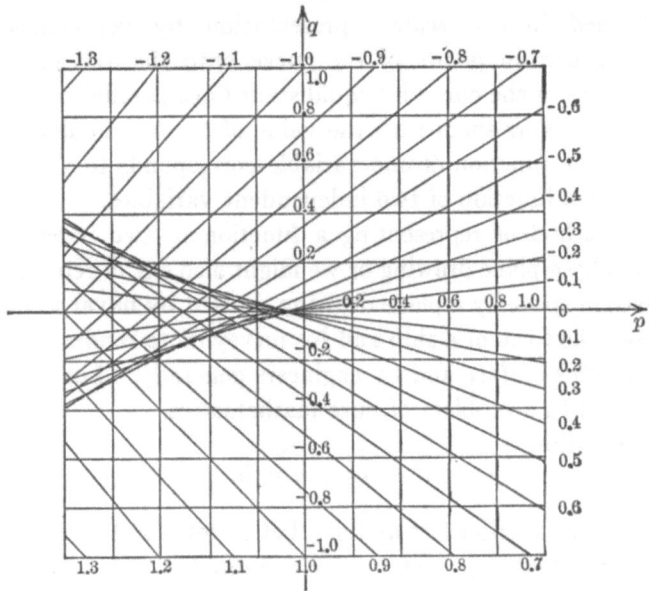

FIG. 44.

On the left side of the square there is a triangular-shaped region where the straight lines cross each other. To each point within this region corresponds an equation with three real roots. For example, at the point $p = -0.8$ and $q = +0.2$ we read $z = -1.00$; $+0.28$; $+0.72$. On the border of this region two roots coincide.

For values of p and q beyond the limits -1 to $+1$ the diagram may also be used. We only have to introduce $z' = z/m$ instead of z and to choose m sufficiently large.

Instead of

$$z^3 + pz + q = 0$$

we obtain

$$m^3 z'^3 + pmz' + q = 0,$$

or dividing by m^3,

$$z'^3 + \frac{p}{m^2} z' + \frac{q}{m^3} = 0,$$

or

$$z'^3 + p'z' + q' = 0,$$

where

$$p' = \frac{p}{m^2}, \quad q' = \frac{q}{m^3}.$$

By choosing a sufficiently large value of m, p' and q' can be made to lie within the limits -1 to $+1$ so that the roots z' may be read on the diagram. Multiplying them by m we obtain the roots z of the given equation.

A function of two independent variables need not be expressed in an explicit form, but may be given in the form of an equation between three variables

$$g(u, v, w) = 0,$$

and we may consider any two of them as independent and the third as a function of the two. The graphical representation may sometimes be greatly facilitated by modifying the method described before. The curves for constant values of one of the three variables, say w, are not plotted by taking u and v as rectangular coördinates, but they are plotted after introducing new variables x and y, x a function of u and y a function of v and making x and y the rectangular coördinates.

In some cases, for instance, we can succeed by a right choice of the functions $x = \varphi(u)$ and $y = \psi(v)$ in getting straight lines for the curves $w = $ const. This will evidently be the case, when the equation $g(u, v, w) = 0$ can be brought into the form

$$a(w)\varphi(u) + b(w)\psi(v) + c(w) = 0,$$

a, b, c being any functions of w, φ any function of u and ψ any function of v.

For introducing

$$x = \varphi(u), \quad y = \psi(v)$$

the equation will become

$$ax + by + c = 0,$$

where a, b, c are constants for any constant value of w.

As an example let us consider the relation between the true solar time, the height of the sun over the horizon, and the declination of the sun for a place of given latitude. Instead of the declination of the sun we might also substitute the time of the year, as the time of the year is determined by the declination of the sun. Our object then is to make a diagram for a place of given latitude from which for any time of the year and any height of the sun the true solar time may be read.

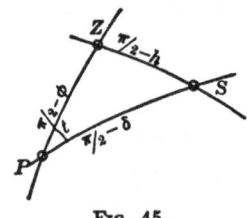

FIG. 45.

In the spherical triangle formed by the zenith Z, the north pole P (if we suppose the place to be on the northern hemisphere) and the sun S (Fig. 45), the sides are the complements of the declination δ, the height h, and the latitude φ. The angle t at the pole is the hour angle of the sun, which expressed in time gives true solar time.

The equation between these four quantities may be written in the form

$$\sin h = \sin \varphi \sin \delta + \cos \varphi \cos \delta \cos t.$$

The latitude φ is to be kept constant, so that t, h, δ are the only variables.

Now let us write

$$x = \cos t, \quad y = \sin h,$$

so that the equation takes the form

$$y = \sin \varphi \sin \delta + x \cos \varphi \cos \delta.$$

When x and y are plotted as rectangular coördinates, we obtain a straight line for any value of δ. Let us draw horizontal lines for equidistant values of $h = 0$ to $90°$ and vertical lines for equidistant values of $t = -180°$ to $+180°$ or expressed in time from midnight to midnight (Fig. 46). In order to draw the

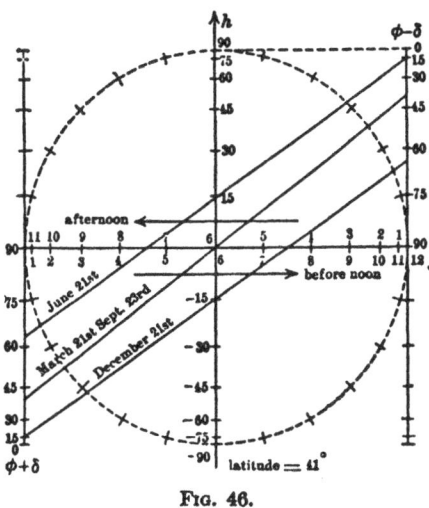

FIG. 46.

straight lines $\delta = $ const., let us calculate where they intersect the vertical lines corresponding to $x = -1$ and $x = +1$ or expressed in time corresponding to midnight and to noon. For $x = -1$ we have $y = -\cos(\varphi + \delta)$, and for $x = +1$ we have $y = \cos(\varphi - \delta)$. Let us draw a scale on the vertical $x = -1$ showing the points $y = -\cos(\varphi + \delta)$ for equidistant values of $(\varphi + \delta)$ and a scale on the vertical $x = +1$, showing the points $y = \cos(\varphi - \delta)$ for equidistant values of $\varphi - \delta$. The scale is the same as the scale for h, with the sole difference that the values of $\varphi - \delta$ are the complements of h and the values of $\varphi + \delta$ the complements of $-h$. For a latitude of $41°$, for instance, we have

For	δ	$\varphi + \delta$	$\varphi - \delta$
June 21.........................	23.5°	64.5°	17.5°
September 23 and March 21........	0	41°	41°
December 21....................	−23.5°	17.5°	64.5°

The values of $\varphi + \delta$ and $\varphi - \delta$ furnish the intersections with
the verticals $x = -1$ and $x = +1$, so that the straight lines
can be drawn corresponding to these days of the year. The two
outward lines are parallel but the middle line is steeper. Their
intersections with the horizontal line $h = 0$ show the time of
sunrise and sunset.[1] Strictly speaking the straight lines do
not correspond to certain days. The straight line determined
by any value of δ changes its position continually as δ changes
continually. But the changes of δ during one day are scarcely
appreciable unless the drawing is on a larger scale.

If in the equation

$$ax + by + c = 0$$

a and b are independent of w, only c being a function of w, all
the straight lines $w = $ const. are parallel. In this case we are
not obliged to draw the
straight lines $w = $ const.
It will suffice to draw a
line perpendicular to the
lines $w = $ const. and a
scale on it that marks the
points corresponding to
equidistant values of w.
On the drawing we place a
sheet of transparent paper
or celluloid, on which three
straight lines are drawn is-
suing from one point in the direction perpendicular to the u-scale,
v-scale and w-scale (Fig. 47). If we move the transparent material
without turning it and make the first two lines intersect the u-and-v
scale at given points, the w-scale will be intersected at the point
corresponding to the value of w. This method has the advantage

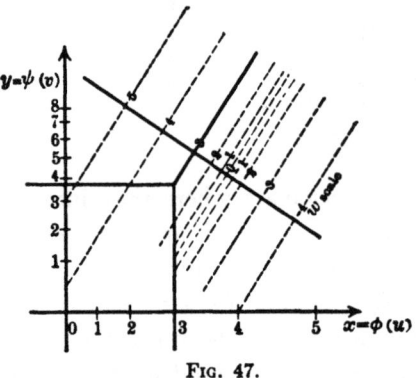

Fig. 47.

[1] That is to say, the moment when the center of the sun would be seen on
the horizon, if there were no atmospherical refraction. To take account of
the refraction, the line $h = -0.6°$ would have to be considered instead of
$h = 0$.

that we can use the same paper for a great many relations of three variables, as we can place a great many scales side by side. Or, in the case of one relation only, we may divide the region of the values u, v, w into a number of smaller regions and draw three scales for each of them, placing all the u-scales or v-scales or w-scales side by side. The drawing will then have the same accuracy as a drawing of very much larger size in which there is only one scale for each of the three variables.

§ 10. *Depiction of One Plane on Another Plane.*—Let us now consider two quantities x and y each as a function of two other quantities u and v

$$x = \varphi(u, v),$$

$$y = \psi(u, v).$$

In order to give a geometrical meaning to this relation between two pairs of quantities let us consider x and y as rectangular coördinates of a point in a plane and u, v as rectangular coördinates of a point in another plane. We then have a correspondence between the two points. When the functions $\varphi(u, v)$ and $\psi(u, v)$ are defined for the values u, v of a certain region, they will furnish for every point u, v of this region a point in the xy plane. Let us call this a depiction of the uv plane on the xy plane. Similarly a function of one variable $x = \varphi(u)$ might be said to depict the u line on the x line. We may therefore say that the depiction of one plane on another plane is, in a certain way, the generalization of the idea of a function of one variable. Let us suppose $\varphi(u, v)$ and $\psi(u, v)$ both to have only one value for given values of u and v for which they are defined. Then there will be only one point in the xy plane corresponding to a given point in the uv plane. But to a given point in the xy plane there may very well correspond several points in the uv plane.

Let us try to explain this by a graphical representation of the depiction of planes on each other. For this purpose we draw the curves $x =$ const. and $y =$ const. in the uv plane for equidistant values of x and y. In the xy plane they correspond to equidistant lines parallel to the axis of x and to the axis of y. The point of intersection of two lines $x = a$ and $y = b$ corresponds to the points of intersection of the curves

$$\varphi(u, v) = a \quad \text{and} \quad \psi(u, v) = b,$$

in the uv plane. If in a certain region of the uv plane, that we consider, they intersect only once there is only one point in the region of the uv plane considered and one point in the xy plane corresponding to each other. Fig. 48 shows the depiction of part of the uv plane on part of the xy plane. We have a net of square-shaped meshes in the xy plane and corresponding is a net of curvilinear meshes in the uv plane.

Let us consider the curves $x = $ const. in the uv plane as the perpendicular projections of curves of equal height on a surface extended over that part of the uv plane. From any point P of the surface corresponding to the values u, v we proceed an

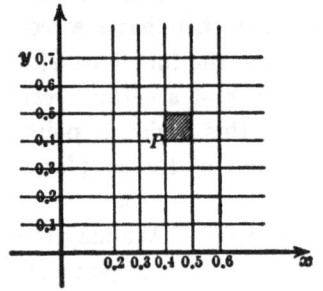 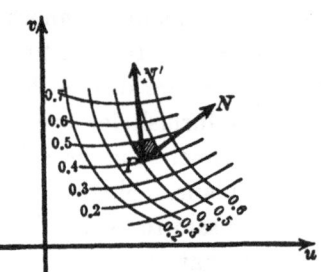

FIG. 48.

infinitely small distance, u changing to $u + du$, v to $v + dv$ and x to $x + dx$, where

$$dx = \frac{\partial \varphi}{\partial u} du + \frac{\partial \varphi}{\partial v} dv.$$

Let us write

$$du = \cos \alpha \, ds, \quad dv = \sin \alpha \, ds,$$

where ds signifies the length of the infinitely small line from u, v to $u + du$, $v + dv$ in the uv plane and α the angle its direction forms with the positive axis of x. Let PN be a straight line whose projections on the u and v axis are equal to $\partial \varphi / \partial u$ and $\partial \varphi / \partial v$ and let us write

$$\frac{\partial \varphi}{\partial u} = r \cos \lambda, \quad \frac{\partial \varphi}{\partial v} = r \sin \lambda,$$

r being the positive length of PN and λ the angle between its direction and the positive axis of x. Then we have

$$dx = \frac{\partial \varphi}{\partial u} du + \frac{\partial \varphi}{\partial v} dv = r ds \cos (\alpha - \lambda),$$

or

$$\frac{dx}{ds} = r \cos (\alpha - \lambda).$$

dx/ds measures the steepness of the ascent. It is positive when the direction leads upward and negative when it leads downward and its value is equal to the tangent of the angle of the ascent. From the equation

$$\frac{dx}{ds} = r \cos (\alpha - \lambda)$$

we see that the ascent is steepest for $\alpha = \lambda$, where $dx/ds = r$. The line PN in the u, v-plane shows the perpendicular projection of the line of steepest ascent on the surface $x = \varphi(u, v)$ and the length of PN measured in the same unit of length in which u and v are measured is equal to the tangent of the angle of the ascent. Let us call the line PN the gradient of the function $\varphi(u, v)$ at the point u, v. The direction of the gradient is perpendicular to the curve $\varphi(u, v) =$ const. that passes through the point u, v; for in the direction of the curve we have

$$\frac{dx}{ds} = 0,$$

and therefore

$$\alpha - \lambda = \pm 90°.$$

If PN' is the gradient of the function $\psi(u, v)$ at the point u, v, the angle between PN and PN' must be equal to the angle formed by the curves x = const. and y = const. that intersect at the point u, v, or equal to its supplement according to the angle of intersection that we consider.

Suppose the gradients PN and PN' do not vanish in any of the points in the region of the uv plane that we consider and that their length and direction vary as continuous functions of u and v. Let us further suppose that the gradient PN' (components: $\partial\psi/\partial u$, $\partial\psi/\partial v$) is for the whole region on the left side of the gradient PN (components: $\partial\varphi/\partial u$, $\partial\varphi/\partial v$), or else for the whole region on the right side of the gradient PN, then it follows that any one of the curves x = const. and any one of the curves y = const. can only intersect once in the region considered.

This may be shown by considering the directions of the curves x = const. and y = const. in the uv plane. Let us consider *that* direction on the curve y = const. in which x increases. If this direction deviates from PN the deviation must be less than 90°, because dx/ds and therefore cos $(\alpha - \lambda)$ is positive. Let us further consider that direction on the curve x = const. in which y increases. If it deviates from the direction of PN' the deviation must be less than 90°. Let us call these directions the direction of x (on the curve y = const.) and the direction of y (on the curve x = const.). Now if the gradient PN' is on the left of the gradient PN the y direction must also be on the left of PN (for if it were on the right of PN being perpendicular to PN it would form an obtuse angle with PN') and therefore it must be on the left of the x direction (for if it were on the right, PN' being perpendicular to the x direction would form an obtuse angle with the y direction, which we have seen to be impossible). Similarly it may be seen, that if PN' is on the right of PN, the direction of y will also be on the right of the direction of x. If therefore PN' is on the same side of PN in the whole region considered, the direction of y will also be on the same side of the direction of x for the whole region considered. This excludes the intersection of two curves x = const. and y = const. in more than one point. For, suppose there are two points of intersection and we pass along the curve y = const. in the direction of x. At the first point of intersection we pass over the curve x = const. from the side of smaller values of x to the side of larger values of x. Now if the values of x go on increasing as we go along the curve y = const. we evidently cannot get back to a curve x = const. corresponding to a smaller value of x. The only possibility of a second point of intersection would be that the direction in which the value of x increases on the curve y = const. becomes the opposite, so that in advancing in the same direction in which we came x would decrease again.

The same holds for the curve x = const. If we pass from one point of intersection with a curve y = const. along a curve x = const. to a second point of intersection with the same curve the only possibility is that the direction of y also becomes opposite. This is excluded as in contradiction with the direction of y being on the

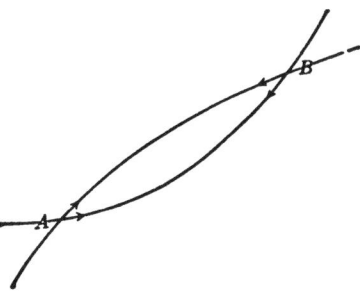

Fig. 49.

same side of the direction of x throughout the whole region (Fig.49)

It will be useful to look at it from another point of view. Let us consider a point A in the uv plane corresponding to the values u, v and let us increase u and v by infinitely small positive amounts du and dv, so that we get four points $ABCD$, forming a rectangle corresponding to the coördinates.

$$A : u, v; \ B : u + du, v; \ C : u, v + dv; \ D : u + du, v + dv.$$

In the xy plane these points are depicted in the points A, B, C, D, the intersections of two curves u and $u + du$ with two curves v and $v + dv$ (Fig. 50).

The projections of the line AB in the xy plane on the axes of coördinates are obtained by calculating the changes of x and y for a constant value of v and a change du in the value of u

$$dx_1 = \frac{\partial \varphi}{\partial u} du, \quad dy_1 = \frac{\partial \psi}{\partial u} du.$$

Similarly the projections of AC are obtained by calculating the changes of x and y for a constant value of u and a change dv in the value of v

$$dx_2 = \frac{\partial \varphi}{dv} dv, \quad dy_2 = \frac{\partial \psi}{\partial v} dv.$$

Denoting the lengths of AB and AC by ds_1 and ds_2 and the angles that the directions of AB and AC form with the direction of the

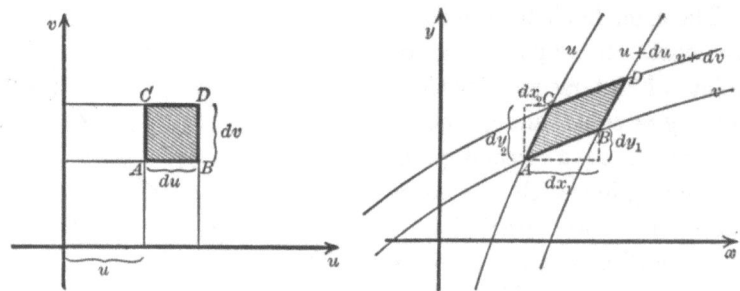

FIG. 50.

positive axis of x (the angles counted in the usual way) by γ_1 and γ_2 we have:

$$dx_1 = ds_1 \cos \gamma_1, \quad dy_1 = ds_1 \sin \gamma_1$$

and

$$dx_2 = ds_2 \cos \gamma_2, \quad dy_2 = ds_2 \sin \gamma_2,$$

or

$$\frac{\partial \varphi}{\partial u} = \cos \gamma_1 \frac{ds_1}{du}, \quad \frac{\partial \psi}{\partial u} = \sin \gamma_1 \frac{ds_1}{du}$$

and

$$\frac{\partial \varphi}{\partial v} = \cos \gamma_2 \frac{ds_2}{dv}, \quad \frac{\partial \psi}{\partial v} = \sin \gamma_2 \frac{ds_2}{dv}.$$

We may call

$$\frac{ds_1}{du} = \sqrt{\left(\frac{\partial \varphi}{\partial u}\right)^2 + \left(\frac{\partial \psi}{\partial u}\right)^2}$$

the scale of depiction at A in the direction AB and

$$\frac{ds_2}{dv} = \sqrt{\left(\frac{\partial \varphi}{\partial v}\right)^2 + \left(\frac{\partial \psi}{\partial v}\right)^2}$$

the scale of depiction at A in the direction AC. It is here understood that the uv plane is the original, which is depicted on the xy plane. If we take it the other way the scales of depiction in the directions AB and AC are the reciprocal values du/ds_1 and dv/ds_2.

The area of the parallelogram $ABCD$ in the xy plane is

$$ds_1 ds_2 \sin (\gamma_2 - \gamma_1) = \left(\frac{\partial \varphi}{\partial u}\frac{\partial \psi}{\partial v} - \frac{\partial \varphi}{\partial v}\frac{\partial \psi}{\partial u}\right) du dv.$$

According to the way in which the angles γ_2 and γ_1 are defined $\sin (\gamma_2 - \gamma_1)$ is positive, when the direction AC points to the left of the direction AB (assuming the positive axis of y to the left of the positive axis of x), and $\sin (\gamma_2 - \gamma_1)$ is negative, when AC points to the right. Now $dudv$ is equal to the area of the rectangle $ABCD$ in the uv plane. Therefore the value of

$$\frac{\partial \varphi}{\partial u}\frac{\partial \psi}{\partial v} - \frac{\partial \varphi}{\partial v}\frac{\partial \psi}{\partial u}$$

is the ratio of the areas $ABCD$ in the two planes and its positive or negative sign denotes the relative position of the directions AB and AC in the xy plane. We may call this ratio the scale of depiction of areas at the point A.

$$\frac{\partial \varphi}{\partial u}\frac{\partial \psi}{\partial v} - \frac{\partial \varphi}{\partial v}\frac{\partial \psi}{\partial u}$$

is called the functional determinant of the functions $\varphi(u, v)$ and $\psi(u, v)$.

We have found the scale of depiction of lengths in the directions AB and AC. Let us now try to find it in any direction whatever. From any point A in the uv plane, whose coördinates are u and v, we pass to a point D close by whose coördinates are $u + \Delta u$, $v + \Delta v$. In the xy plane we find the corresponding points A and D with coördinates (Fig. 51).

$$A: \begin{array}{l} x = \varphi(u, v) \\ y = \psi(u, v) \end{array} \quad D: \begin{array}{l} x + \Delta x = \varphi(u + \Delta u, v + \Delta v) \\ y + \Delta y = \psi(u + \Delta u, v + \Delta v) \end{array}$$

We expand according to Taylor's theorem, and writing for shortness

$$\varphi_u = \frac{\partial \varphi}{\partial u}, \quad \varphi_v = \frac{\partial \varphi}{\partial v}, \quad \psi_u = \frac{\partial \psi}{\partial u}, \quad \psi_v = \frac{\partial \psi}{\partial v}$$

we find

$$\Delta x = \varphi_u \Delta u + \varphi_v \Delta v + \text{terms of higher order,}$$

$$\Delta y = \psi_u \Delta u + \psi_v \Delta v + \text{terms of higher order.}$$

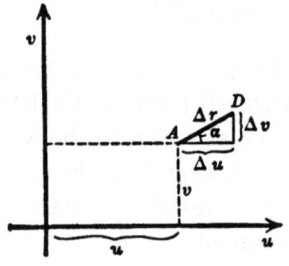 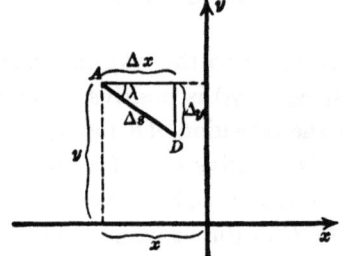

FIG. 51.

The length of AD and the angle of its direction we denote by Δr and α in the uv plane and by Δs and λ in the xy plane. The limit of the ratio $\Delta s/\Delta r$, to which it tends, when D approaches A without changing the direction AD is the scale of depiction at the point A in the direction AD.

Writing

$$\Delta u = \Delta r \cos \alpha,$$
$$\Delta v = \Delta r \sin \alpha,$$

we obtain

$$\Delta x = (\varphi_u \cos \alpha + \varphi_v \sin \alpha)\Delta r + \text{terms of higher order,}$$

$$\Delta y = (\psi_u \cos \alpha + \psi_v \sin \alpha)\Delta r + \text{terms of higher order.}$$

Dividing by Δr and letting Δr decrease indefinitely, we have in the limit

$$\frac{dx}{dr} = \varphi_u \cos \alpha + \varphi_v \sin \alpha,$$

$$\frac{dy}{dr} = \psi_u \cos \alpha + \psi_v \sin \alpha.$$

For dx/dr and dy/dr we may also write $ds/dr \cos \lambda$, $ds/dr \sin \lambda$.

$$\frac{ds}{dr} \cos \lambda = \varphi_u \cos \alpha + \varphi_v \sin \alpha,$$

$$\frac{ds}{dr} \sin \lambda = \psi_u \cos \alpha + \psi_v \sin \alpha.$$

These equations show the scale of depiction ds/dr corresponding to the different directions λ in the x, y-plane and α in the u, v-plane.

By introducing complex numbers we can show the connection still better.

Let us denote

$$z = \frac{dx}{dr} + \frac{dy}{dr} i = \frac{ds}{dr} e^{\lambda i},$$

$$z_1 = \varphi_u + \psi_u i,$$

$$z_2 = \varphi_v + \psi_v i.$$

Multiplying the second of the two equations by i and adding both they may be written as one equation in the complex form:

$$z = z_1 \cos \alpha + z_2 \sin \alpha.$$

The modulus of z is the scale of depiction of the uv plane at the point A in the direction α. The angle of z gives the direction in the xy plane corresponding to the direction α. For $\alpha = 0$ we have $z = z_1$ and for $\alpha = 90°$, $z = z_2$.

Let us substitute

$$\cos \alpha = \frac{e^{\alpha i} + e^{-\alpha i}}{2}, \quad \sin \alpha = \frac{e^{\alpha i} - e^{-\alpha i}}{2i}$$

and write

$$a = \frac{z_1 + z_2/i}{2}, \quad b = \frac{z_1 - z_2/i}{2},$$

so that the expression for z becomes

$$z = ae^{\alpha i} + be^{-\alpha i}.$$

This suggests a simple geometrical construction of the complex numbers z for different values of α. The term $ae^{\alpha i}$ is represented by the points of a circle described by turning the line that represents the complex number a round the origin through the angles $\alpha = 0 \cdots 2\pi$. The term $be^{-\alpha i}$ is represented by the points of a circle described by turning the line that represents b round the origin in the opposite direction through the angles $\alpha = 0 \cdots -2\pi$ (Fig. 52). The addition of the two complex numbers $ae^{\alpha i}$ and $be^{\alpha i}$ for any value of α is easily performed. The points corresponding to the complex numbers z describe an ellipse, whose two principal axes bisect the angles between a and b. This is easily seen by writing

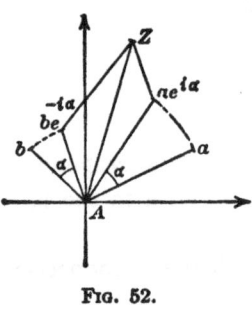

FIG. 52.

$$a = r_1 e^{(\alpha_0 - \alpha_1)i}, \quad b = r_2 e^{(\alpha_0 + \alpha_1)i}.$$

α_0 corresponds to the direction bisecting the angle between a and b and α_1 denotes half the angle between a and b (positive or negative according to the position of a and b).

$$z = r_1 e^{(\alpha_0 - \alpha_1 + \alpha)i} + r_2 e^{(\alpha_0 + \alpha_1 - \alpha)i},$$

or

$$ze^{-\alpha_0 i} = r_1 e^{(\alpha - \alpha_1)i} + r_2 e^{-(\alpha - \alpha_1)i}$$
$$= (r_1 + r_2) \cos (\alpha - \alpha_1) + (r_1 - r_2) \sin (\alpha - \alpha_1)i.$$

Denoting the coördinates of the complex number $ze^{-\alpha_0 i}$ by ξ and η we have

$$\frac{\xi}{r_1 + r_2} = \cos (\alpha - \alpha_1) \quad \text{and} \quad \frac{\eta}{r_1 - r_2} = \sin (\alpha - \alpha_1),$$

and consequently the equation of an ellipse

$$\frac{\xi^2}{(r_1 + r_2)^2} + \frac{\eta^2}{(r_1 - r_2)^2} = 1.$$

This ellipse turned round the origin through an angle equal to α_0 gives us the points corresponding to z. The principal axes are $2(r_1 + r_2)$ and $2(r_1 - r_2)$ (Fig. 53). The construction of

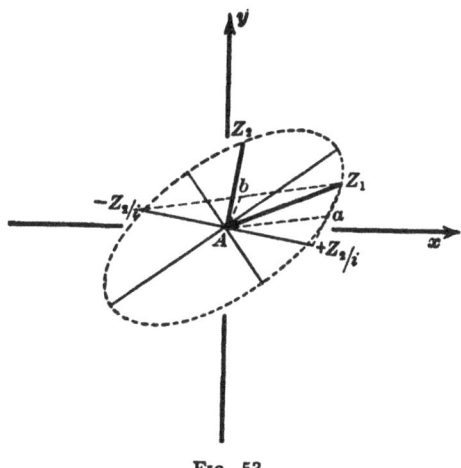

FIG. 53.

Fig. 53 is obvious. After plotting z_1 and z_2 we find z_2/i and $- z_2/i$ by turning AZ_2 through a right angle to the right and to the left. From these points lines are drawn to Z_1. The bisection of these lines give a and b.

The figure shows that in case a and b have the same modulus, the triangle $- Z_2/i$, Z_1, Z_2/i becomes equilateral and AZ_1 is perpendicular to the line joining $- Z_2/i$ and Z_2/i. In this case AZ_1 and AZ_2 would have the same or the opposite direction. But as $z_1 = \varphi_u + \psi_u i$, $z_2 = \varphi_v + \psi_v i$, this would mean that $\varphi_u \psi_v - \varphi_v \psi_u = 0$.

The radii of the ellipse (Fig. 53) measured in the unit used give the different scales of depiction corresponding to the different directions in the xy plane. We might also say the ellipse is the image in the xy plane of an infinitely small circle in the uv plane, magnified in the proportion of the infinitely small radius to 1, with its center in A.

Z_1 corresponds to $\alpha = 0$ and Z_2 to $\alpha = 90°$ and for $\alpha = 0$ to $90°$

Z moves on the ellipse from Z_1 to Z_2 through the shorter way.
$- Z_1$ corresponds to $\alpha = 180°$ and $- Z_2$ to $\alpha = 270°$. Now we
have shown above that a positive value of the functional deter-
minant $\varphi_u \psi_v - \varphi_v \psi_u$ means that Z_2 is on the positive side of Z_1,
so that in this case Z moves in the positive sense (that is, in the
direction from the positive axis of x to the positive axis of y) with
increasing values of α. With a negative value Z moves in the
opposite direction.

Let us now suppose that the curves $x = $ const. and $y = $ const. in
the uv plane intersect except on a certain curve where their direc-

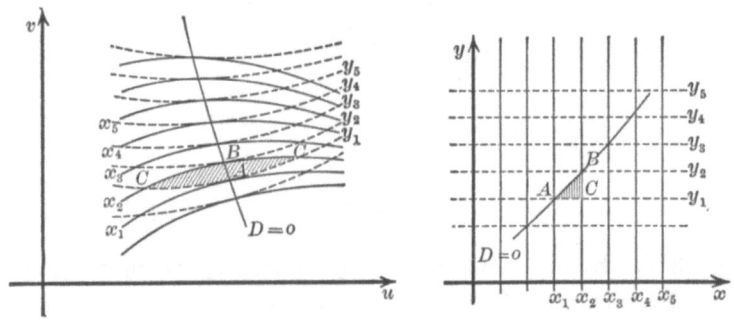

FIG. 54.

tions coincide in the way shown in Fig. 54. On this curve the
functional determinant $D = \varphi_u \psi_v - \varphi_v \psi_u$ must vanish because
the directions of the gradients coincide. Let us see what the
depiction on the xy plane is like.

Running along one of the curves $y = $ const., say $y = y_1$,
toward the curve $D = 0$ we intersect the curves $x = x_4, x_3, x_2$
until at the point A on the curve $x = x_1$ we reach the curve $D = 0$.
In the xy plane the corresponding path is a parallel to the axis
of x at a distance y_1 passing through x_4, x_3, x_2 and reaching a
point A at x_1. If we now proceed on the curve $y = y_1$ in the
uv plane beyond the curve $D = 0$, we again intersect the curves
$x_2, x_3,$ etc., but in the inverse order. Thus the corresponding
path in the xy plane does not pass beyond A, but turns back

through the same points x_2, y_1; x_3, y_1, etc. The same holds for any of the other lines $y =$ const. If we trace the line in the xy plane that corresponds to the points in the uv plane, where the curves $x =$ const. and $y =$ const. touch, we find the depiction of the uv plane only on one side of the curve in the xy plane. The other side has no corresponding points u, v. However to every point C on this side of the curve, there are two corresponding points C in the uv plane, one on either side of the curve $D = 0$. Imagine two sheets of paper laid on the xy plane; let them both be cut along the curve AB. Retain only the two pieces on this side of the curve and paste them together along the curve. The uv plane is in this way depicted on the paper in such a way that there is one point and one only on the paper corresponding to each point in the region of the uv plane considered. The curve $D = 0$ in the uv plane corresponds to the rim where the two pieces of paper are pasted together. Any line straight or curved passing over the curve $D = 0$ in the uv

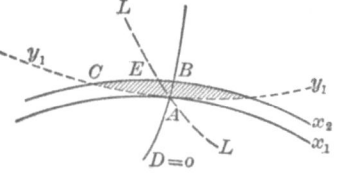

Fig. 55.

plane, corresponds to a line running from one of the sheets onto the other. It need not change its direction abruptly when it reaches the rim and passes onto the other sheet. For it may touch the rim in the direction of its tangent. This is actually the rule and the abrupt change of direction is the exception. Any line LAL (Fig. 55) in the uv plane, whose tangent as it crosses the curve $D = 0$ at A does not coincide with the common tangent of the curves $x =$ const. and $y =$ const. will correspond to a line in the xy plane, that does not change its direction abruptly when it touches the rim.

This is best understood analytically. Let us consider corresponding directions at the points A in the uv plane and in the xy plane. We have seen above that corresponding directions (Fig. 56) are connected by the equations

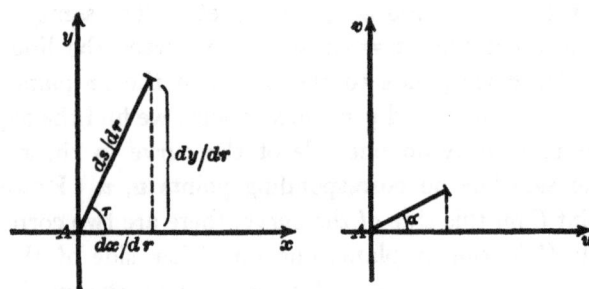

Fig. 56.

$$\cos \lambda \frac{ds}{dr} = \frac{dx}{dr} = \varphi_u \cos \alpha + \varphi_v \sin \alpha,$$

$$\sin \lambda \frac{ds}{dr} = \frac{dy}{dr} = \psi_u \cos \alpha + \psi_v \sin \alpha.$$

At the point A we have

$$\varphi_u \psi_v - \varphi_v \psi_u = 0.$$

Assuming that the gradients at A do not vanish, so that we can write

$$\varphi_u = r \cos \gamma, \quad \varphi_v = r \sin \gamma,$$

$$\psi_u = r' \cos \gamma', \quad \psi_v = r' \sin \gamma',$$

where r and r' are positive quantities, the equation $\varphi_u \psi_v - \varphi_v \psi_u = 0$ reduces to $\sin (\gamma - \gamma') = 0$, that is, $\gamma = \gamma'$ or $\gamma = \gamma' + 180°$. It follows therefore that:

$$\cos \lambda \frac{ds}{dr} = r \cos (\alpha - \gamma),$$

$$\sin \lambda \frac{ds}{dr} = r' \cos (\alpha - \gamma') = \pm r' \cos (\alpha - \gamma).$$

Consequently for all directions α in the uv plane for which $\cos (\alpha - \gamma)$ is not zero, we have

$$\operatorname{tg} \lambda = \pm \frac{r'}{r}.$$

That is to say, we have in the xy plane only one fixed direction λ and the opposite corresponding to all the different directions α except only a direction for which $\cos(\alpha - \gamma) = 0$.   In the latter case, that is, when the direction α is perpendicular to the direction γ of the gradient, $i.\ e.$, in the direction of the curves $x = $ const. and $y = $ const., we have

$$\cos\lambda\frac{ds}{dr} = 0,$$

$$\sin\lambda\frac{ds}{dr} = 0.$$

Therefore $ds/dr = 0$ and λ remains indeterminate.   Any direction λ for which $\mathrm{tg}\,\lambda$ differs from $+\ r'/r$ corresponds to a fixed direction $\alpha = \gamma + 90°$ or $\alpha = \gamma - 90°$, while $ds/dr = 0$.

As the curve $D = 0$ is depicted on the rim of the two sheets of paper, all those lines that intersect the curve $D = 0$ in a direction different from the direction of the curves $x = $ const. and $y = $ const. are depicted in the xy plane as curves having their tangent at A in common with the rim.   All lines in one of the sheets of paper that touch the rim at A in a direction different from that of the rim must be the depiction of lines in the uv plane that reach A in the direction of the lines $x = $ const. and $y = $ const.   The scale of depiction is zero in the direction of the curves $x = $ const. and $y = $ const.   In any other direction α we find it different from zero for:

$$\frac{ds}{dr} = \sqrt{(r^2 + r'^2)\cos^2(\alpha - \gamma)}.$$

It is a maximum in the direction $\alpha = \gamma$ or $\gamma + 180°$ perpendicular to the curves $x = $ const. and $y = $ const.

It may help to understand all these details if we discuss an example where the depiction of the uv plane on the xy plane has a simple geometrical meaning, the planes being ground plan and elevation of a curved surface in space.   The rim in the xy plane is the outline of the surface, the projection of those points where the tangential plane is perpendicular to the plane of elevation.

Suppose a cylinder of circular section cut in two half cylinders by a plane through its axis. Suppose one of the half cylinders

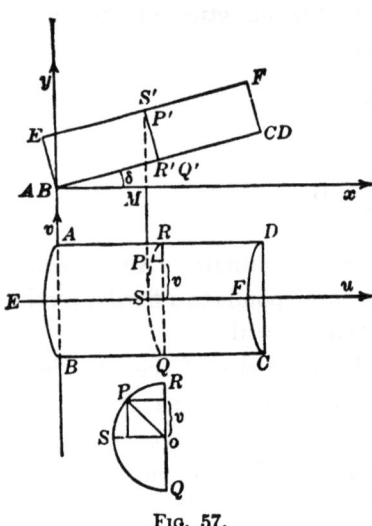

FIG. 57.

in such a position that its axis forms an angle δ with the ground plan, the plan of elevation being parallel to its axis, Fig. 57. Let us introduce rectangular coördinates u, v in the ground plan and rectangular coördinates x, y in the plan of elevation. A point P on the cylinder is defined by certain values u, v which define its ground plan and certain values x, y which define its elevation. It is easily seen from Fig. 57 that we have

$$x = u$$

and

$$y = u \, \mathrm{tg}\delta + \frac{1}{\cos \delta} \sqrt{a^2 - v^2},$$

where a is the radius of the section. Now let us consider the elevation of the points P as a depiction of their ground plan. The functions $\varphi(u, v)$ and $\psi(u, v)$ in this case are

$$\varphi(u, \ v) = u,$$
$$\psi(u, v) = u \, \mathrm{tg} \, \delta + \frac{1}{\cos \delta} \sqrt{a^2 - v^2},$$

and

$$\varphi_u = 1, \quad \varphi_v = 0; \quad \psi_u = \mathrm{tg} \, \delta, \quad \psi_v = - \frac{v}{\cos \delta \sqrt{a^2 - v^2}},$$

$$\varphi_u \psi_v - \varphi_v \psi_u = - \frac{v}{\cos \delta \sqrt{a^2 - v^2}}.$$

The functional determinant vanishes for $v = 0$ on the line EF. The lines $y = $ const. are the intersections of the cylinder with horizontal planes. In the plan of elevation they are straight horizontal lines; in the ground plan they are ellipses (Fig. 58). As we pass along one of these curves we cross the line EF in the ground plan but we only touch it in the plan of elevation, retracing the horizontal line back again. The lines $x = $ const. are straight lines in both planes, but in space they correspond to ellipses. Again as we cross EF in the ground plan we only touch it in the plan of elevation and retrace the vertical line down again. Any curve on the cylinder that crosses EF in a direction not perpendicular to the plan of elevation is projected in the plan of elevation with EF as its tangent. For the real tangent in space lying in the tangential plane of the cylinder can have no other projection, if not perpendicular to the plan of elevation. In this latter case the projection of the tangent is a point and the tangent of the elevation is determined by the inclination of the osculatory plane.

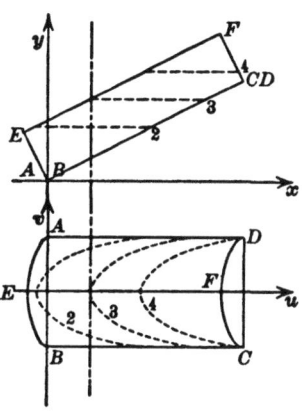

Fig. 58.

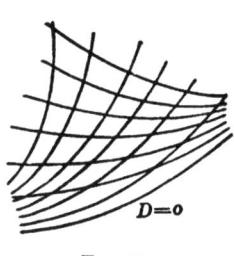

Fig. 59.

There is a particular case to be considered, when the curve $D = 0$ in the uv plane coincides with one of the curves $x = $ const. or $y = $ const. (Fig. 59), assuming the gradients of the functions $\varphi(u, v)$ and $\psi(u, v)$ not to vanish at the points of this curve. We have seen that at a point where $D = 0$ the scale of depiction must vanish in the directions of the curve $x = $ const. or $y = $ const. Let the curve $D = 0$ coincide with a line $x = $ const., then it follows that the

7

length of the depiction of this curve is zero and the depiction must be contracted in a point. For the length of the depiction of a curve $x = $ const. is given by an integral

$$\int \frac{ds}{dr}\, dr,$$

where dr denotes an element of the curve and ds/dr the scale of depiction in the direction of the curve. As ds/dr is zero all along the curve the integral must necessarily vanish.

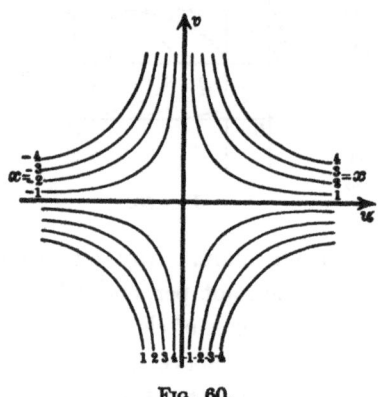

As an example let us consider

$$x = uv,$$

$$y = v.$$

The lines $x = $ const. in the uv plane are equilateral hyperbolas, the lines $y = $ const. are parallels to the axis of u (Fig. 60). Along the axis of u we have at the same time $y = 0$, $x = 0$ and $D = v = 0$. The whole axis of u is depicted in

Fig. 60.

the point $x = 0$, $y = 0$ of the xy plane.

Let us finally consider the case where the scale of depiction at any point is the same in all directions, though it need not be the same at different points.

Writing as before

$$z_1 = \varphi_u + \psi_u i, \quad z_2 = \varphi_v + \psi_v i,$$

$$z = \frac{dx}{dr} + \frac{dy}{dr} i = \frac{ds}{dr} e^{\lambda i},$$

the connection between the scale of depiction ds/dr and the angles λ, α determining corresponding directions in the xy plane and in the uv plane is given by the equation

$$z = z_1 \cos \alpha + z_2 \sin \alpha,$$

or

$$z = ae^{ia} + be^{-ia},$$

where

$$a = \tfrac{1}{2}(z_1 + z_2/i), \quad b = \tfrac{1}{2}(z_1 - z_2/i).$$

In the case where the scale of depiction ds/dr, that is to say, the modulus of z, is independent of α, one of the constants a or b must vanish, as we see at once from the construction of z (Fig. 52). Let us consider the case $b = 0$,

$$z = ae^{ia} = \frac{ds}{dr} e^{i\lambda}.$$

The complex number a may be written $|a|\ e^{ia_0}$, where $|a|$ denotes the modulus of a and α_0 the angle. Both may vary from point to point, but at every point they have fixed values.

Consequently we have

$$\frac{ds}{dr} = |a| \quad \text{and} \quad \lambda = \alpha + \alpha_0.$$

That is to say, from an angle α determining a direction in the uv plane, we find the angle λ determining the corresponding direction in the xy plane by the addition of a fixed value α_0. Any two directions α, α' will therefore form the same angle as the corresponding directions λ, λ' in the xy plane. The same is true when $a = 0$ and $z = be^{-ia}$. The only difference is that in this latter case the direction of z rotates in the opposite sense with increasing values of α.

Analytically depictions of this kind are represented by functions of complex numbers,

$$x + yi = f(u + vi) \quad \text{or} \quad x + yi = f(u - vi).$$

Assuming the function to possess a differential coefficient we have

$$z_1 = \frac{\partial x}{\partial u} + \frac{\partial y}{\partial u}i = f'(u \pm vi),$$

$$z_2 = \frac{\partial x}{\partial v} + \frac{\partial y}{\partial v}i = \pm f'(u \pm vi)i,$$

and therefore either

$$z_1 = z_2/i \quad \text{or} \quad z_1 = - z_2/i.$$

Hence in the first case

$$a = \tfrac{1}{2}(z_1 + z_2/i) = z_1, \quad b = \tfrac{1}{2}(z_1 - z_2/i) = 0$$

and in the second case

$$a = 0, \quad b = z_1.$$

§ 11. *Other Methods of Representing Relations between Three Variables.*—The depiction of one plane on another may be used to generalize the graphical representation of a function of two variables or a relation between three variables, as we prefer to say.

As we have seen before, an equation

$$g(x, y, z) = 0$$

between three variables x, y, z can be represented by taking x and y as rectangular coördinates and plotting the curves $z =$

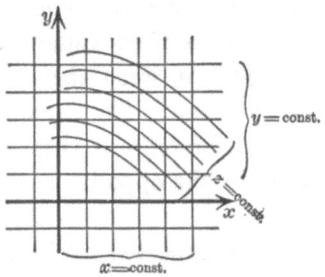

$y = \text{const.}$

$z = \text{const.}$

$x = \text{const.}$

FIG. 61.

const. (Fig. 61) for equidistant values of z. Suppose now the xy plane to be depicted on another plane. The lines $x = \text{const.}, y = \text{const.}$ and $z = \text{const.}$ will be represented by three sets of curves. The fact that three values x, y, z satisfy the equation $g(x, y, z) = 0$ is shown geometrically by the intersection of the three corresponding curves in one point.

Another method for representing certain relations between three variables u, v, w consists in drawing three curves, each curve carrying a scale. The values of u, v, w are read each on one of the three scales. The relation between three values u, v, w is represented geometrically by the condition that the corresponding points lie on a straight line (Fig. 62). This method is far more convenient than the one using three sets of curves. It is less trouble to place a ruler over two points u, v of two curves and read the value w on the scale of the third than to find the intersection of two curves u = const. and v = const. among sets of others, pick out the curve w = const. that passes through the

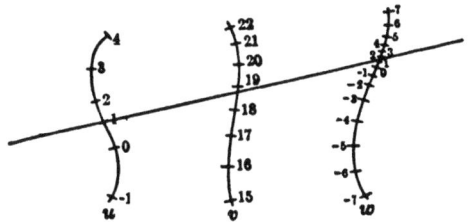

FIG. 62.

same point and read the value of w corresponding to it. For we must consider that the curves corresponding to certain values of u and v are generally not drawn, but must be interpolated and so must the curve w = const. It is true that interpolations are necessary with both methods, but the interpolation on scales like those in Fig. 62 is easily done.

It must however be understood that while the three sets of curves form a perfectly general method for representing any relation between three variables, the other method is restricted to certain cases. In order to investigate this subject more fully we shall have to explain the use of line coördinates.

When we apply rectangular coördinates x, y to define a certain point in a plane, we may say that x determines one of a set of straight lines (parallel to the axis of ordinates) and y determines one of another set of straight lines (parallel to the axis of abscissas) and the point is the intersection of the two (Fig. 63, I). A similar method may be used to determine a certain straight line in a plane. Let x determine a *point* on a certain straight line, x being its distance from a fixed point A on the line measured in a certain unit and counted positive on one side and negative on the other. Let y define a *point* on another straight line

parallel to the first, y being its distance from a fixed point B on
the line measured in the same way as x. The straight line
passing through the two points is thus determined by the values

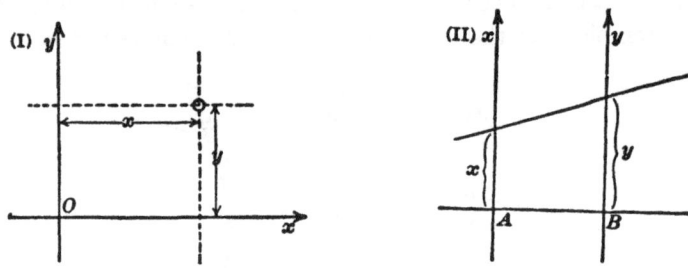

FIG. 63.

x and y and for all possible values of x and y we obtain all the
straight lines of the plane except those parallel to the lines on
which x and y are measured. For simplicity we choose AB
perpendicular to the two lines (Fig. 63, *II*). Let us call x and y
the line coördinates of the line connecting the two points x and
y in Fig. 63, *II*, in the same way as x and y in Fig. 63, *I*, are
called the point coördinates of the point where the two lines
x and y intersect.

A linear equation between point coördinates

$$y = mx + \mu$$

is the equation of a straight line. That is to say, all the points
whose coördinates satisfy the equation lie on a certain straight
line. If, on the other hand, we regard x and y as line coördinates
we find the analogous theorem: all the straight lines whose
line coördinates satisfy the equation

$$y = mx + \mu$$

pass through a certain point. The equation is therefore called
the equation of the point.

In order to show this let us first draw the line $x = 0$, $y = \mu$
(APO in Fig. 64). If now for any value of x we make $AR = x$

and $PQ = mx$, the point of intersection of RQ and AP must be independent of x, for

$$\frac{PO}{AO} = \frac{mx}{x} = m.$$

The ratio PO/AO determines the position of O and as it is independent of x and the positions of A and P are also independent of x, the same is true for O.

For negative values of m, PO and AO have opposite directions so that O lies between A and P.

For a given point O, we can find the corresponding values of m and μ by joining O with the points A and the point corresponding to $x = 1$. If P and Q are the intersections of these lines with the line on which y

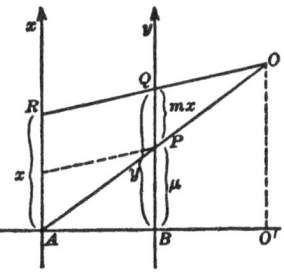

FIG. 64.

is measured, we have $BP = \mu$ and $PQ = m$. Any point in the plane thus leads to an equation

$$y = mx + \mu,$$

except the points on the line on which x is measured. For $m = 0$ the equation reduces to

$$y = \mu,$$

that is, the equation of a point on the line on which y is measured.

Instead of $y = mx + \mu$, we might also write $x = m'y + \mu'$, and go through similar considerations changing the parts of x and y. This form does not include the points on the line on which y is measured, but it does include the points on the line on which x is measured. For these we have $m' = 0$.

The general equation of a point in line coördinates is given in the form

$$ax + by + c = 0,$$

from which we may derive either of the first-mentioned forms dividing it by a or b.

Dividing by c another convenient form is obtained,

$$\frac{ax}{-c} + \frac{by}{-c} = 1,$$

or writing

$$\frac{-c}{a} = x_0, \quad \frac{-c}{b} = y_0,$$

$$\frac{x}{x_0} + \frac{y}{y_0} = 1,$$

x_0 determining the point of intersection of the line BO (Fig. 64) and the x-line, while y_0 determines the point of intersection of the line AO with the y-line.

A curve may be given by an equation

$$a_1(u)x + b_1(u)y + c_1(u) = 0,$$

in which $a_1(u)$, $b_1(u)$, $c_1(u)$ are functions of a variable u. Any value of u furnishes the equation of a certain point and as u changes the point describes the curve. Let us suppose the curve drawn and a scale marked on it giving the values of u in certain intervals sufficiently close to interpolate the values of u between them. Two other curves are in the same way given by the equations

$$a_2(v)x + b_2(v)y + c_2(v) = 0,$$

$$a_3(w)x + b_3(w)y + c_3(w) = 0,$$

and scales on these curves mark the values of v and w.

Now we are enabled to formulate the condition which must be satisfied by the values u, v, w in order that the three corresponding points lie in one straight line. If x and y are the line coördinates of the line passing through the three points, x and y must satisfy all three equations simultaneously.

Consequently the determinant of the three equations must vanish

$$a_1(b_2c_3 - b_3c_2) + a_2(b_3c_1 - b_1c_3) + a_3(b_1c_2 - b_2c_1) = 0,$$

and, vice versa, if the equation between u, v, w may be brought into this form where a_1, b_1, c_1 are any functions of u, a_2, b_2, c_2 any functions of v and a_3, b_3, c_3 any functions of w, we can form the equations

$$a_1x + b_1y + c_1 = 0,$$
$$a_2x + b_2y + c_2 = 0,$$
$$a_3x + b_3y + c_3 = 0,$$

and represent them graphically by curves carrying scales for u, v, w. The relation between u, v, w is then equivalent to the condition that the corresponding points on the three curves lie on a straight line. But it must be remembered that only a restricted class of relations can be brought into the required form, so that the method cannot be applied to any given relation.

The equation of a point

$$ax + by + c = 0$$

remains of the same form, when the units of length are changed for x and y. If x' denotes the number measuring the same length as the number x but in another unit, the two numbers must have a constant ratio equal to the inverse ratio of the two units. Therefore, by changing the units independently, we have

$$x = \lambda x', \quad y = \mu y',$$

and the equation of the point may be written

$$a\lambda x' + b\mu y' + c = 0,$$

or

$$a'x' + b'y' + c = 0,$$

where $a' = \lambda a$ and $b' = \mu b$.

It is sometimes convenient to define the line coördinates in another way. Let ξ and η denote rectangular coördinates measured in the same unit, then the equation of a straight line can be written

$$\eta = \text{tg } \varphi \xi + \eta_0,$$

where φ is the angle between the line and the axis of ξ and η_0,

the ordinate of the point of intersection with the axis of η.
Now let us call tg φ and η_0 the line coördinates of the straight
line represented by the equation and let us denote them by x
and y. Thus the values of x and y define a certain straight line
and any straight line not parallel to the axis of ordinates may
be defined in this manner. The condition that a straight line
x, y passes through a point ξ, η is expressed by the equation

$$\eta = x\xi + y,$$

or

$$y = -\xi x + \eta.$$

If we fix the values of x and y, all the values ξ, η that satisfy this
equation represent the points of the straight line x, y and we
therefore call it the equation of the straight line. If, on the
other hand, we fix the values of ξ and η, all the values x, y that
satisfy the equation represent the straight lines that pass through
the given point ξ, η, and therefore we call it the equation of the
point.

The more general form

$$ax + by + c = 0$$

can be reduced to

$$y = -\frac{a}{b}x - \frac{c}{b}.$$

It therefore represents the equation of the point, whose rectangular coördinates are $\xi = a/b$ and $\eta = -c/b$. The case
where $b = 0$ or

$$ax + c = 0$$

represents the equation of a point infinitely far away in the
direction φ or the opposite direction $\varphi + 180°$, φ being defined by

$$\text{tg } \varphi = x = -\frac{c}{a}.$$

All the straight lines, whose coördinates x, y satisfy the equation

$$ax + c = 0$$

correspond to the same value of x but to any value of y. That is to say, they are all parallel and all the straight lines of this direction belong to them.

Let us now discuss some of the applications of line coördinates to the graphical representation of relations between three variables.

The relation

$$uv = w$$

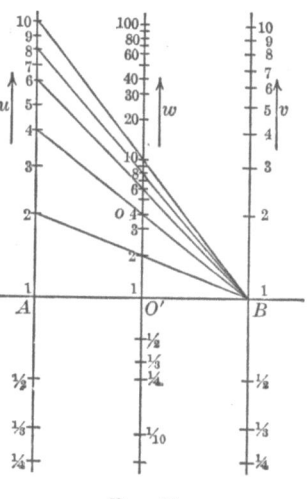

may be written in the form

$$\log u + \log v = \log w,$$
or
$$x + y = \log w,$$
when

$$x = \log u \quad \text{and} \quad y = \log v.$$

Let us plot x and y as line coördinates on two parallel lines (Fig. 65), with scales for the values of u and v. The equations $x = \log u$

and $y = \log v$ may be regarded as the equations of the points of these two scales. The equation

$$x + y = \log w$$

for any value of w is the equation of a point. It can easily be constructed as the intersection of any lines x, y satisfying its equation. For instance, the line $x = \log w$, $y = 0$ and the line $x = 0$, $y = \log w$. The first line is found by connecting the scale division $u = w$ of the u-scale with the point B, the second by connecting the scale division $v = w$ of the v-scale with the point A. If the units of x and y are taken of the same length, the point of intersection will lie in the middle between the two lines carrying the u and v scales on a line parallel to the two other lines and the w-scale will be half the size of the other two (Fig. 65).

The relation

$$uv = w$$

or

$$\log u + \log v = \log w$$

expresses the condition that the three equations

$$x = \log u, \quad y = \log v, \quad x + y = \log w$$

are satisfied simultaneously by the same values of x and y, that is to say, that the three points on the u, v, w scales corresponding to the values of u, v, w lie on the same straight line x, y.

The more general relation

$$u^{\alpha}v^{\beta} = w,$$

where α and β are any given values, can be treated in the same manner. Thus the pressure and volume of a gas undergoing adiabatic changes may be represented. In this case we have

$$pv^k = w,$$

where p denotes the pressure, v the volume and k and w constants.

For a given gas k has a given value, but w depends on the quantity of the gas considered.

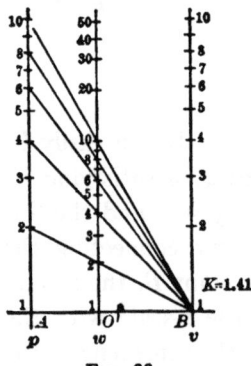

FIG. 66.

We write

$$x = \log p, \quad y = \log v.$$

The relation then takes the form

$$x + ky = \log w,$$

and represents a point which may be constructed by the intersection of any two straight lines x, y, whose coördinates satisfy the equation, for instance

$$x = \log w, \quad y = 0$$

and

$$x = 0, \quad y = \frac{1}{k} \log w.$$

The first line connects the point B (Fig. 66) with the scale division $p = w$ of the p scale and the second line connects the point A with the scale division of the v scale for which $y = k \log w$. A perpendicular from the point of intersection on AB meets it in O' and as the ratio $AO'/O'B$ is equal to the ratio of the segments on the p and v scales $\log w/k \log w = 1/k$ it is independent of w. All the points corresponding to different values of w lie on the same parallel to the p and v scales and the w scale may be obtained by a central projection of the p scale on this parallel from the center B (Fig. 66). We might dispense with the construction of the w scale as long as the straight line for the w scale is drawn. For in using the diagram we generally start with values p_0, v_0 and want to find other values p, v, for which

$$pv^k = p_0 v_0^k.$$

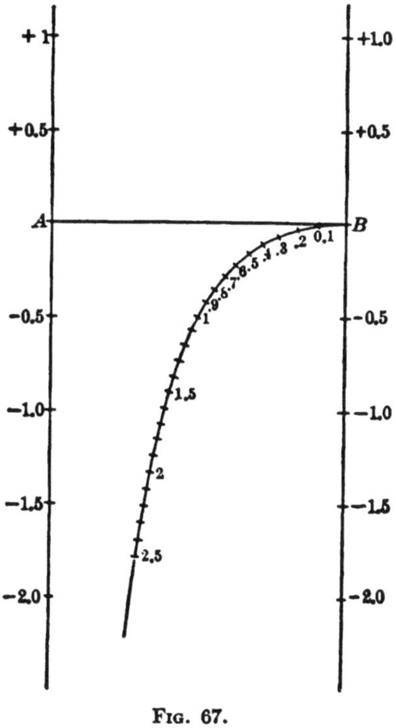

FIG. 67.

The straight line connecting the scale divisions p and v intersects the w scale at the same point as the straight line connecting the scale divisions p_0 and v_0, so that we need not know the value of $p_0 v_0^k$. It suffices to mark the point of intersection in order to find the value of p, when v is given or the value of v when p is given.

Another example is furnished by the equation

$$w^2 + xw + y = 0.$$

If we regard x and y as line coördinates any value of w determines the equation of a point. We plot the curve formed by these points with a scale on it indicating the corresponding values of w. Any values of x and y determine a straight line whose intersections with the w scale furnish the roots of the equation. Each point of the w scale may be constructed by the intersection of two straight lines, whose coördinates x, y satisfy the equation, for instance

$$x = 0, \quad y = -w^2 \quad \text{and} \quad x = -w, \quad y = 0.[1]$$

In Fig. 67 the w scale is shown for the positive values $w = 0$ to $w = 2.5$.

In the same manner a diagram for the solution of the cubic equation

$$w^3 + xw + y = 0.$$

or of any equation of the form

$$w^\lambda + xw^\mu + y = 0$$

may be constructed.

§ 12. *Relations between Four Variables.*—The method can be generalized for relations between four variables.

Suppose four variables u, v, w, t are connected by the equation

$$g(u, v, w, t) = 0,$$

and let us assume that for any particular value $t = t_0$ the resulting relation between u, v, w can be given by a diagram of the form considered consisting of three curves carrying scales for u, v and w. Let us further suppose that for other values of t the scales for u and v remain the same, but the scale for w changes. Then we shall have a set of w scales corresponding to different values of t. Connecting the points that correspond to the same value of w we obtain a network of curves $t = $ const. and $w = $ const. (Fig. 68). Any two values u, v furnish a straight line intersecting

---

[1] For small values of w, this combination is not good because the angle of intersection is small. One might substitute $x = 2$, $y = -w^3 - 2w$ for the first line.

the network of curves.   The points of intersection correspond to values of t and w that satisfy the given relation.

Any relation of the form

$$\varphi(u)f(t,\,w) + \psi(v)g(t,\,w) + h(t,\,w) = 0$$

may be represented in this way, $\varphi(u)$ denoting any function of u, $\psi(v)$ any function of v and $f(t,$ $w)$, $g(t,\,w)$, $h(t,\,w)$ any functions of t and w.

In this case we need only introduce the line coördinates x, y, writing

$$x = \varphi(u), \quad y = \psi(v).$$

We then obtain a linear equation between x and y,

$$f(t,\,w)x + g(t,\,w)y + h(t,\,w) = 0,$$

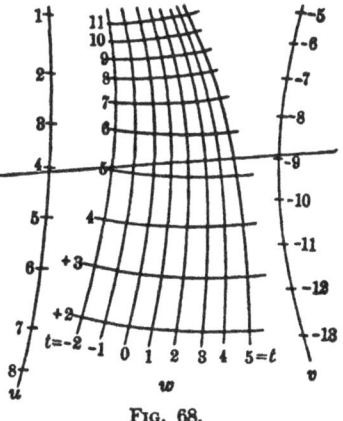

FIG. 68.

which for any given values of t and w represents the equation of a point. For a given value of t and variable values of w we obtain a curve $t = $ const. carrying a scale for w and for a series of values of t we obtain a set of curves $t = $ const. Similarly for a given value of w and variable values of t the equation furnishes a curve $w = $ const., carrying a scale for t and a series of values of w furnishes a set of curves $w = $ const. From any given values of u and v the line coördinates x and y are calculated and the points where this straight line defined by x and y intersects the network of the curves $t = $ const. and $w = $ const. furnish the values t, w that satisfy the relation together with the given values of u and v. The relation between the height, azimuth, declination of a celestial body and the latitude of the point of observation may serve as an example. Let h, a, δ denote the height, azimuth and declination and φ the latitude. The angles $\pi/2 - \varphi$, $\pi/2 - h$, $\pi/2 - \delta$ are the three sides of a spherical triangle PZS (Fig. 69) formed by the pole P, the zenith Z and the celestial body S. The azimuth is defined as the supplement of the angle PZS.

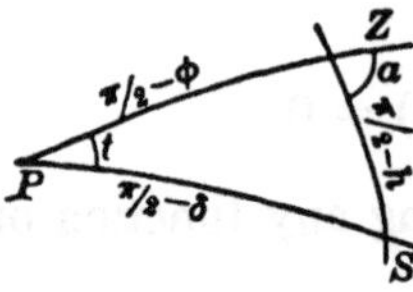

The equation is

$$\sin \delta = \sin \varphi \sin h - \cos \varphi \cos h \cos a.$$

We write

$$x = \cos a, \quad y = \sin \delta,$$

so that the equation becomes

$$y = \sin \varphi \sin h - x \cos \varphi \cos h.$$

Fig. 69.

We shall in this case use the second system of line coördinates where x is the slope of the line measured by the tangent of the angle formed with the axis of abscissas and y is the ordinate of the intersection with the axis of ordinates. If ξ, η denote the rectangular coördinates of the point, the equation of the points takes the form

$$\eta = x\xi + y \quad \text{or} \quad y = \eta - \xi x,$$

so that in our case we have

$$\xi = \cos \varphi \cos h, \quad \eta = \sin \varphi \sin h.$$

The curves $\varphi = $ const. and $h = $ const. can be drawn by means of these formulas. It is easily seen that they are ellipses and that the curves $\varphi = $ const. are the same as the curves $h = $ const. For a definite value of φ and a variable value of h we find

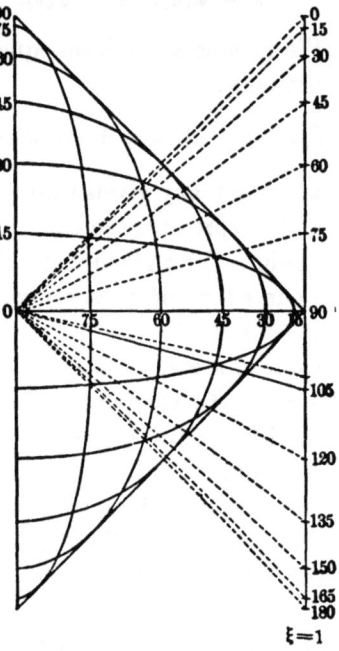

Fig. 70.

$$\frac{\xi^2}{\cos^2 \varphi} + \frac{\eta^2}{\sin^2 \varphi} = 1,$$

and for a definite value of h and a variable value of φ

$$\frac{\xi^2}{\cos^2 h} + \frac{\eta^2}{\sin^2 h} = 1.$$

Any of the ellipses intersects all the others and in this way they form a network. A point of intersection of the ellipse $\varphi = c_1$ and the ellipse $h = c_2$ also corresponds to the values $h = c_1$ and $\varphi = c_2$, as the ellipse $\varphi = c_1$ is identical with the ellipse $h = c_1$ and $\varphi = c_2$ identical with $h = c_2$ (Fig. 70). The easiest way to find this network consists in drawing the straight lines

$$\xi + \eta = \cos (\varphi - h),$$

and perpendicular to them the straight lines

$$\xi - \eta = \cos (\varphi + h),$$

for equidistant values of $\varphi + h$ and $\varphi - h$. The ellipses run diagonally through the rectangular meshes formed by the two systems of straight lines. The scales for φ and h are written on the axis of coördinates, both scales being available for both variables. The scale for δ is written on the axis of ordinates and is identical with the scale for t and h on this axis. For the ordinate corresponding to a given value $\delta = c$ is $\sin c$, and this is also the ordinate of the point where the ellipse $\varphi = c$ or $h = c$ intersects the axis of ordinates. The scale for the azimuth cannot be laid down in exactly the same way as that for φ, h and δ because $\cos a$ determines the slope of the straight line x, y. Let us draw a parallel to the axis of ordinates through the point $\xi = 1$, $\eta = 0$ and mark a scale for the azimuth on it, making $\eta = \cos a$ (Fig. 70). A line connecting the origin with any scale division of this scale has the slope of the line $x = \cos a$, $y = \sin \delta$. To bring it into the position of the line x, y it must be moved parallel to itself, until its point of intersection with the axis of ordinates coincides with the scale division δ. This suggests another way of using the diagram. Let a pencil of rays be drawn from the origin to the scale divisions of the azimuth scale (Fig. 70), and let it be drawn on a sheet of transparent paper

8

placed over the drawing of the ellipses. For any given value of δ it is moved up or down as the case may be so that the center of the pencil coincides with the scale division δ. As long as the celestial body does not materially alter its declination the diagram in this position will enable us to find any of the three values φ, h, a from the other two.

As a second example let us consider the relation between the declination δ, the azimuth a, the hour angle t of a celestial body and the latitude φ of the point of observation.

The relation is found by eliminating the height h from the equation

$$\sin \delta = \sin \varphi \sin h - \cos \varphi \cos h \cos a.$$

For this purpose we express $\sin h$ and $\cos h$ by the other angles and substitute these expressions for $\sin h$ and $\cos h$.

We have

$$\cos h = \cos \delta \sin t / \sin a,$$

$$\sin h = \sin \varphi \sin \delta + \cos \varphi \cos \delta \cos t.$$

Substituting these values we find

$$\sin \delta = \sin^2 \varphi \sin \delta + \sin \varphi \cos \varphi \cos \delta \cos t - \cos \varphi \cos \delta \sin t \operatorname{ctg} a, \text{ or}$$

$$\cos^2 \varphi \sin \delta = \sin \varphi \cos \varphi \cos \delta \cos t - \cos \varphi \cos \delta \sin t \operatorname{ctg} a.$$

Dividing by $\cos^2 \varphi \cos \delta$ we finally obtain

$$\operatorname{tg} \delta = \operatorname{tg} \varphi \cos t - \frac{\sin t}{\cos \varphi} \operatorname{ctg} a.$$

In order to represent this relation graphically we introduce line coördinates

$$x = \operatorname{ctg} a \quad \text{and} \quad y = \operatorname{tg} \delta$$

and find

$$y = \operatorname{tg} \varphi \cos t - \frac{\sin t}{\cos \varphi} x.$$

Let us use the second system of line coördinates. The rectangular coördinates ξ, η of the point represented by the equation are found from it equal to:

$$\xi = \frac{\sin t}{\cos \varphi}, \quad \eta = tg \; \varphi \cos t.$$

The curves $\varphi = $ const. are ellipses,

$$\cos^2 \varphi \xi^2 + \frac{\eta^2}{tg^2 \varphi} = 1.$$

The curves $t = $ const. are hyperbolas,

$$\frac{\xi^2}{\sin^2 t} - \frac{\eta^2}{\cos^2 t} = 1.$$

The ellipses and hyperbolas are confocal, the foci coinciding with the points $\xi = \pm 1$, $\eta = 0$, so that the curves intersect at right angles.

The scale for φ may be written on the axis of ordinates at the points where it intersects the ellipses. It is identical with the scale for δ, the ordinate in both cases being the tangent of the angle with the only difference that δ is negative on the negative part of the axis and φ is not. . The scale for t may be written on one of the ellipses corresponding to the largest value of φ that is to be taken account of. This ellipse forms the boundary of the diagram, so that larger values of φ are not represented. Corresponding to the azimuth we draw a pencil of rays on a sheet of transparent paper, which is laid on the drawing of the curves. The center of the pencil is placed on the scale division δ and the azimuth is equal to the angles

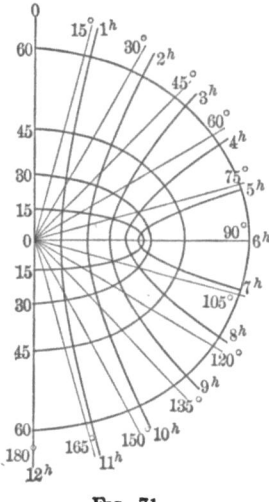

FIG. 71.

that the rays form with the positive direction of the axis of ordinates (Fig. 71). It suffices to draw the curves and the rays only on one side of the axis of ordinates. At the apex of the hyperbolas the value of t changes abruptly.   The line $t = 6^h$ is meant to start from the focus $\xi = 1, \eta = 0$.   When the center of the pencil of rays is in the origin the rays form the asymptotic lines of the hyperbolas, $a = 15°$ corresponding to $t = 1^h$, $a = 30°$ to $t = 2^h$ and so on.

CHAPTER III.

THE GRAPHICAL METHODS OF THE DIFFERENTIAL AND INTEGRAL CALCULUS.

§ 13. *Graphical Integration.*—We have shown how the elementary mathematical operations of adding, subtracting, multiplying and dividing and the inverse operation of finding the root of an equation can be carried out by graphical methods and how functions of one or more variables may be represented and handled. But the graphical methods would lack generality and would be of very limited use, if they were not applicable to the infinitesimal operations of differentiation and integration. Indeed it is here that they are found of the greatest value. In many cases, where the calculus is applied to problems of natural science or of engineering, the functions concerned are given in a graphical form. Their true analytical structure is not known and as a rule an approximation by analytical expressions is not easily calculated nor easily handled. In these cases it is of vital importance that the operations of the calculus can be performed, although the functions are only given graphically.

Let us begin with integration, because it is easier than differentiation and of more general application.

Suppose a function $y = f(x)$ given by a curve whose ordinate is y and whose abscissa is x. The problem is to find a curve, whose ordinate Y is an integral of the function $f(x)$,

$$Y = \int_a^x f(x)dx.$$

Let us assume the unit of length for the abscissas independent of the unit of length for the ordinates. The value of Y measures the area between the ordinates corresponding to a and x, the curve $y = f(x)$ and the axis of x in units equal to the rectangle formed by the units of x and y.

In the simple case where $f(x)$ is a constant the equation $y = f(x) = c$ is represented by a line parallel to the axis of x and

$$Y = \int_a^x c\,dx = c(x - a).$$

Y is the ordinate of a straight line intersecting the axis of x at the point $x = a$. The constant c is the change of Y for an increase of x equal to 1. If P is the point on the axis of x for $x = -1$ and Q the point where the line $y = c$ intersects the axis of ordinates (Fig. 72) the desired line is parallel to PQ. It is constructed by drawing a parallel to PQ through the point $x = a$ on the axis of x (Fig. 72, where $a = 0$).

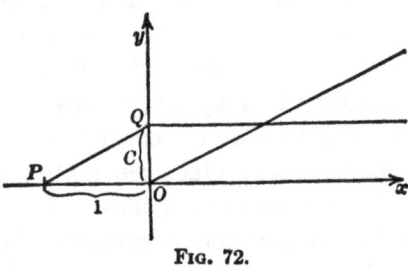

FIG. 72.

When a given value c_1 is added, so that the equation becomes

$$Y = c\,(x - a) + c_1$$

it amounts to the same as when the straight line is moved in the direction of the axis of ordinates through a distance c_1. For $x = a$ we then have $Y = c_1$, so that we obtain the line

$$Y = c(x - a) + c_1,$$

by drawing a parallel to PQ through the point $x = a$, $y = c_1$.

In the second place let us assume that the line $y = f(x)$ consists of a number of steps, that is to say, that the function has different constant values in a number of intervals $x = x_1$ to x_2, x_2 to x_3, etc., while it changes its value abruptly at x_2, x_3, etc. The line presenting the integral

$$Y = \int_{x_1}^x f(x)\,dx$$

does not change its ordinate abruptly. It consists of a continuous broken line, whose corners have the abscissas x_2, x_3, etc.

The directions of the different parts are found in the way just described by the pencil of rays from P to the points α, β, γ, etc. (Fig. 73), where the horizontal lines intersect the axis of ordinates.

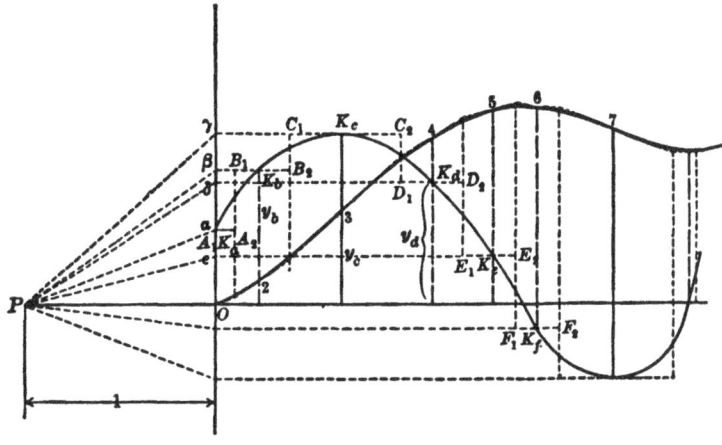

Fig. 73.

To construct the broken line we draw a parallel to $P\alpha$ through the point $x = x_1$ (in Fig. 73 x_1 is equal to 0) as far as the vertical $x = x_2$. Through the point of intersection with the vertical $x = x_2$ we draw a parallel to $P\beta$ as far as the vertical $x = x_3$. Through the point of intersection with the vertical $x = x_3$ we draw a parallel to $P\gamma$ and so on.

Finally let us consider the case of an arbitrary function $y = f(x)$ represented by any curve. In order to find the curve

$$Y = \int_a^x f(x)dx$$

we substitute for $y = f(x)$ a function consisting of different constant values in different intervals and changing its value abruptly when x passes from one interval to the next, so that the line representing this function consists of a number of steps leading up or down according to the increase or decrease of $f(x)$. These steps are arranged in the following way. The horizontal part A_1A_2 of the first step (Fig. 73) starts from any point A_1 of the given curve. The vertical part A_2B_1 and the following horizontal part B_1B_2 are then drawn in such a manner that B_1B_2 intersects the curve and that the integral of the given function as far as the point of intersection K_b is equal to the integral of the stepping line as far as the same point. That is to say, the areas between the stepping line and the given curve on both sides of the vertical part A_2B_1 have to be equal. When K_b is fixed the right position of A_2B_1 may be found by eye estimate. The eye is rather sensitive for differences of small areas. Besides a shift of A_2B_1 to the right or to the left enlarges one area and diminishes the other so that even a slight deviation from the correct position makes itself felt. In the same way the next step $B_2C_1C_2$ is drawn with its vertical part B_2C_1 in such a position that the areas on both sides are equal. The integral of the given curve as far as K_c will again have the same value as that of the stepping line as far as K_c. And so on for the other steps. The integral of the stepping line is constructed in the way shown. It is represented by a broken line beginning at the foot of the ordinate of A_1. The corners lie on the vertical parts of the steps or their prolongations. It is readily seen that the broken line consists of a series of tangents of the integral curve

$$Y = \int_a^x f(x)dx,[1]$$

and that their points of contact with the integral curve lie on the same verticals as the points A_1, K_b, K_c, etc. (In Fig. 73 these points are denoted 0, 2, 3, $\cdots$.) That these points lie on the integral curve follows from the arrangement of the steps which make the integral of the given function at K_b, K_c, $\cdots$ equal to the integral of the stepping line. Now in the points A_1, K_b, K_c $\cdots$ the ordinates of the given curve coincide with those of the stepping line. Hence both integral lines must for these abscissas have the same direction.

[1] In Fig. 73 the lower limit is 0.

Having constructed the broken line and marked the points 2, 3, 4, $\cdots$ (Fig. 73), the integral curve is drawn with a curved ruler so as to touch the broken line in the points, 0, 2, 3, $\cdots$. As the given curve does not change its ordinate abruptly the integral curve does not change its direction abruptly. The drawing shows how well the integral curve is determined by the broken line. There is practically no choice in drawing it any other way without violating the conditions.

The ordinate of the integral curve is measured in the same unit as the ordinate of the given curve $y = f(x)$. It may sometimes be convenient to draw the ordinates of the integral curve in a scale different from that of the ordinates of the given curve. For instance the value of the integral may become so large that measured in the same unit the ordinates of the integral curve would pass the boundaries of the drawing board, or else they may be so small that their changes cannot be measured with sufficient accuracy. In the first case the scale is diminished, in the latter case it is enlarged. This is done by altering the position of the point P, the center of the pencil of rays that define the directions of the broken line. If P approaches O the directions $P\alpha$, $P\beta$, $\cdots$ become steeper to the same degree as if keeping P unchanged we had increased the ordinates of A_1A_2, B_1B_2, $\cdots$ in the inverse proportion of the two distances PO. Hence by diminishing the distance PO the ordinates of the resulting broken line are enlarged in the inverse proportion. On the other hand, by increasing the distance PO the ordinates of the resulting broken line are diminished in the inverse proportion of the distances, because the change of the directions $P\alpha$, $P\beta$, $\cdots$ caused by a longer distance PO is the same as if the ordinates of A_1A_2, B_1B_2, $\cdots$ were diminished in the inverse proportion. The broken line constructed by means of the longer distance $P'O$ will therefore be the same as if the ordinates of the stepping line were diminished. It therefore leads to an integral curve whose ordinates are diminished in the same proportion (Fig. 74).

The graphical integration of

$$Y = \int_a^x f(x)dx$$

is not limited to values $x > a$. The method is just as well applicable to the continuation of the integral curve for $x < a$. The

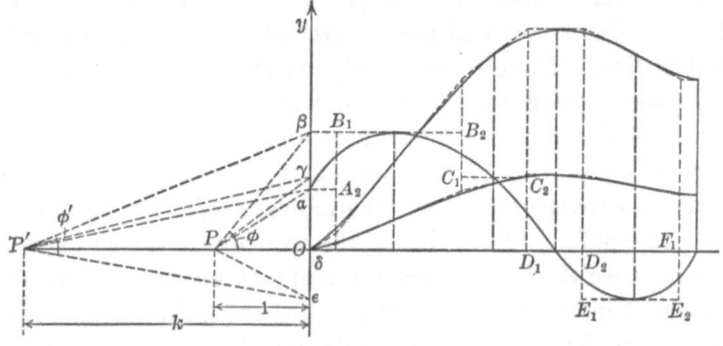

FIG. 74.

steps have only to be drawn from right to left. The lower limit a determines the point where the integral curve intersects the axis of x.

There is a method for the construction of the vertical parts of the steps, which may in some cases be useful, though as a rule we may dispense with it and fix their position by estimation.

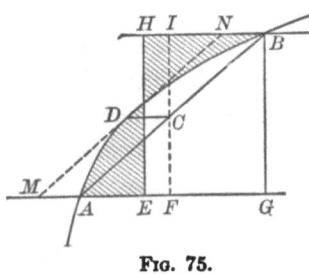

FIG. 75.

Suppose that A and B (Fig. 75) are two points where the curve is intersected by the horizontal parts of two consecutive steps and that the curve between A and B is a parabola whose axis is parallel to the axis of x. The position of the vertical part of the step between A and B can be then found by a simple construction. Through the center C of the chord AB (Fig. 75) draw a parallel CD to the axis of x, D being the point of intersection with the parabola. The vertical part EH of the step intersects CD in a point whose distance from C is twice the distance from D. That this is the right position of EH is shown as soon as we can prove that the area $ADBGA$ is equal to the rectangle $EHBG$. The area $ADGBA$ can be divided in two parts, the triangle ABG and the part $ADBCA$ between the curve and the chord. The triangle is equal to the rectangle $FIBG$, while $ADBCA$ is equal to two thirds of the parallelogram $MNBA$, and hence equal to the rectangle $EHIF$. Both together are therefore equal to the rectangle $EHBG$, and the two areas between the stepping line and the curve on both sides of EH are thus equal.

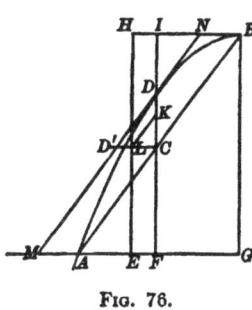

FIG. 76.

If the curve between A and B is supposed to be a parabola with its axis parallel to the axis of ordinates the construction has to be modified a little. Through the center C of the chord AB (Fig. 76) draw a vertical line CD as far as the parabola. On CD find the point K whose distance from C is double the distance from D and draw through it a parallel to the chord AB. This parallel intersects a horizontal line through C at a point L. Then EH must pass through L. This may be shown in the following way. The area between the parabola ADB and the chord AB is equal to two thirds of the parallelogram $MNBA$, MN being the tangent to the parabola at the point D. If D' is the point of intersection of NN and the horizontal line through C, we have evidently

$$CL = \tfrac{2}{3}CD'.$$

Therefore the rectangle $EHIF$ is equal to the area $ADBA$ between the parabola and the chord and $EHBG$ is equal to $ADGBA$.

Any part of a curve can be approximated by the arc of a parabola with sufficient accuracy if the part to be approximated is sufficiently small. When the direction of the curve is nowhere parallel to the axis of coördinates, both kinds of parabolas may be used for approximation, those whose axes are parallel to the axis of x and those whose axes are parallel to the axis of y. But when the direction in one of the points is horizontal (Fig. 76), we can only use those with vertical axes and when the direction in one of the points is vertical we can only use those with horizontal axes. Accordingly we have to use either of the two constructions to find the position of the vertical part of the step.

Do not draw your steps too small. For, although the difference between the broken line and the integral curve becomes smaller, the drawing is liable to an accumulation of small errors owing to the considerable number of corners of the broken line and little errors of drawing committed at the corners. Only practical experience enables one to find the size best adapted to the method.

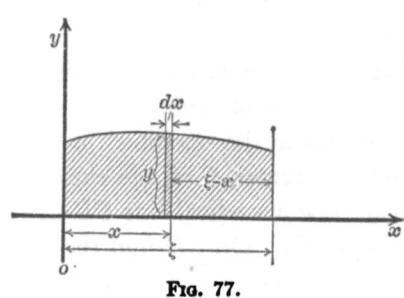

Fig. 77.

Statical moments of areas may be found by a double graphical integration. Let us consider the area between the curve $y = f(x)$ (Fig. 77), the axis of x and the ordinates corresponding to $x = 0$ and $x = \xi$. The statical moment with respect to the vertical through $x = \xi$ is the integral of the products of each element ydx and its distance $\xi - x$ from the vertical

$$M = \int_0^\xi (\xi - x)ydx.$$

Let us regard M as a function of ξ and differentiate it:

$$\frac{dM}{d\xi} = ((\xi - x)y)_{x=\xi} + \int_0^\xi \frac{d}{d\xi}(\xi - x)ydx$$
$$= 0 + \int_0 ydx.$$

That is to say, a graphical integration of the curve $y = f(x)$ beginning at $x = 0$ furnishes the curve whose ordinate is

$$\frac{dM}{d\xi}.$$

Hence a second integration of this latter curve will furnish the curve M as a function of ξ. As M vanishes for $\xi = 0$ the second integration must also begin at the abscissa $x = 0$.

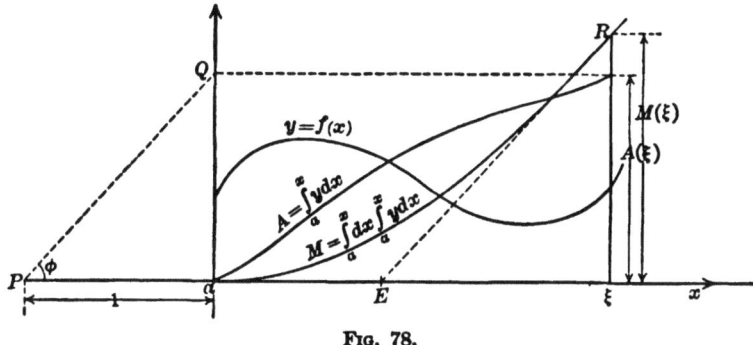

FIG. 78.

Fig. 78 shows an example. Each ordinate of the curve found by the second integration is the statical moment of the area on the left side of it with respect to the vertical through this same ordinate. The ordinate furthest to the right is the statical moment of the whole area with respect to the vertical on the right. The statical moment of the whole area with respect to a vertical line through any point x_1 is the integral

$$\int_0^\xi (x_1 - x)y dx.$$

Considered as a function of x_1 its differential coefficient is

$$\int_0^\xi \frac{d}{dx_1} (x_1 - x)y dx = \int_0^\xi y dx.$$

That is to say, the differential coefficient is independent of x_1, hence the statical moment is represented by a straight line. As its differential coefficient is represented by a horizontal line through the last point on the right of the curve

$$\int_0^\xi y dx,$$

the direction of the straight line is found by drawing a line through P and through the point of intersection Q of the horizontal line and the axis of ordinates (Fig. 78). The position of the straight line is then determined by the condition that

$$\int_0^\xi (x_1 - x)y\,dx$$

for $x_1 = \xi$ is equal to the statical moment

$$M(\xi) = \int_0^\xi (\xi - x)y\,dx.$$

We have therefore only to draw a parallel to PQ through the last point R of the curve for $M(\xi)$ found by the second integration. The ordinates of this straight line for any abscissa x_1 represent the values of

$$\int_0^\xi (x_1 - x)y\,dx$$

measured in the unit of length of the ordinates. The point of intersection E with the axis of x determines the position of the vertical in regard to which the statical moment is zero, that is to say, the vertical through the center of gravity.

The moment of inertia of the area

$$\int_0^\xi y\,dx$$

about the axis $x = \xi$ is found in a similar way. It is expressed by the integral

$$T = \int_0^\xi (\xi - x)^2 y\,dx.$$

Considered as a function of ξ we find by differentiation

$$\frac{dT}{d\xi} = [(\xi - x)^2 y]_{x=\xi} + \int_0^\xi \frac{d}{d\xi}(\xi - x)^2 y\,dx$$

$$= 0 + 2\int_0^\xi (\xi - x)y\,dx.$$

That is to say, the differential coefficient is equal to double the statical moment about the same axis. This holds for every value of ξ. Hence we obtain $\frac{1}{2}T$ as a function of ξ by integrating the curve for $M(\xi)$. For $\xi = 0$ we have $T = 0$, so that the curve begins on the axis of x at $\xi = 0$.

The integral

$$\int_a^x y\,dx$$

is zero for $x = a$. The curve representing the integral has to intersect the axis of x at $x = a$ (admitting values of $x > a$ and $x < a$), and it is there that we begin the construction of the broken line. If instead we begin it at the point $x = a$, $y = c$, the only difference is that the whole integral curve is shifted parallel to the axis of ordinates by an amount equal to c upwards if c is positive, downwards if it is negative. But the form of the curve remains the same. It is different when this curve is integrated a second time. For instead of

$$\int_a^x y\,dx$$

we now integrate

$$\int_a^x y\,dx + c.$$

The ordinate of the integral curve is therefore changed by an amount equal to $c(x - a)$ and besides if the second integral curve is begun at $x = a$, $y = c_1$ instead of $x = a$, $y = 0$ the change amounts to

$$c(x - a) + c_1,$$

so that the difference between the ordinates of the new integral curve and the ordinates of the straight line

$$y = c(x - a) + c_1$$

is equal to the ordinates of the first integral curve (Fig. 79).

This effect of adding a linear function to the ordinates of the integral curve is also attained by shifting the pole P upward or

downward. For it evidently comes to the same thing whether
the curve to be integrated is shifted upward by the amount c or
whether the point P is moved downward by the same amount, so
that the relative position of P and the curve to be integrated
is the same as before. Changing the ordinate of P by $-c$ adds

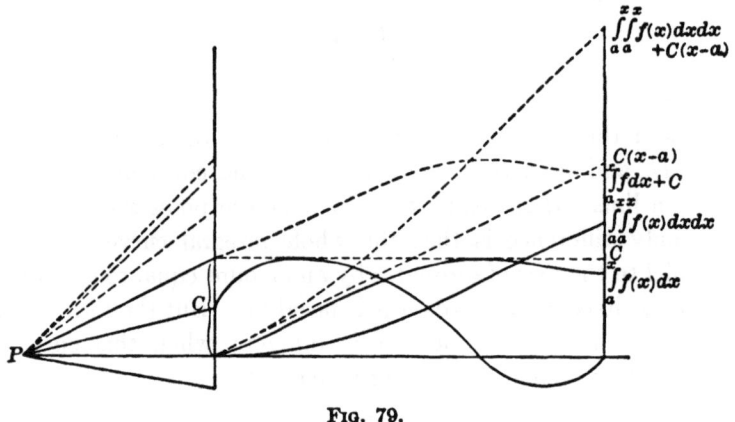

$$\int_a^x \int_a^x f(x)dxdx$$
$$+ C(x-a)$$

$$C(x-a)$$
$$\int_a^x \int f dx + C$$
$$\int_a^x \int_a^x f(x)dxdx$$
$$C$$
$$\int_a^x f(x)dx$$

FIG. 79.

$c(x - a)$ to the ordinates of the integral curve. $c(x - a)$ is the
ordinate of a straight line parallel to the straight line from the
new position of P to the origin.

By this device of shifting the position of P upward or downward the integral curve may sometimes be kept within the
boundaries of the drawing without any reduction of the scale of
ordinates. A good rule is to choose the ordinate of P about
equal to the mean ordinate of the curve to be integrated. The
ordinates of the integral curve will then be nearly the same at
both ends. The value of the integral

$$\int_a^x y\,dx$$

is equal to the difference between the ordinates of the integral
curve and the ordinates of a straight line parallel to PO through
the point of the integral curve whose abscissa is a.

When the ordinate of P is accurately equal to the mean ordinate of the curve to be integrated for the interval $x = a$ to b the ordinates of the integral curve will be accurately the same at the two ends. But we do not know the mean ordinate before having integrated the curve.

After having integrated we find the mean ordinate for the interval $x = a$ to b by drawing a straight line through P parallel to the chord AB of the integral curve, A and B belonging to the abscissas $x=a$ and $x=b$. This line intersects the axis of ordinates at a point whose ordinate is the mean ordinate.

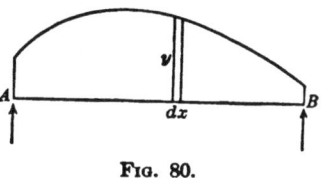

FIG. 80.

Suppose a beam AB is supported at both ends and loaded by a load distributed over the beam as indicated by Fig. 80. That is to say, the load on dx is measured by the area ydx. Let us integrate this curve graphically, beginning at the point A with P on the line AB. The final ordinate at B

$$\int_a ydx$$

gives the whole load and is therefore equal to the sum of the two reactions at A and B that equilibrate the load. Integrating this curve again we obtain the curve whose ordinate is equal to

$$\int_a^{\xi} Ydx,$$

Y being written for

$$\int_a ydx$$

The ordinate of this curve at any point $x = \xi$ represents the statical moment of the load between the verticals $x = a$ and $x = \xi$ about the axis $x = \xi$. Its final ordinate BM, Fig. 81, is the moment of the whole load about the point B, and as the reactions equilibrate the load it must be equal to the moment of the

9

reactions about the same point and therefore opposite to the
moment of the reaction at A about B. If the reaction at A is
denoted by F_a we therefore have

$$F_a(b - a) = \int_a^b Y dx.$$

That is to say, F_a is equal to the mean ordinate of the curve

$$Y = \int_a^\xi y dx$$

in the interval $x = a$ to b. The mean ordinate is found by
drawing a parallel to AM through P which intersects the vertical
through A at the point F so that $AF = F_a$. As DB is equal to

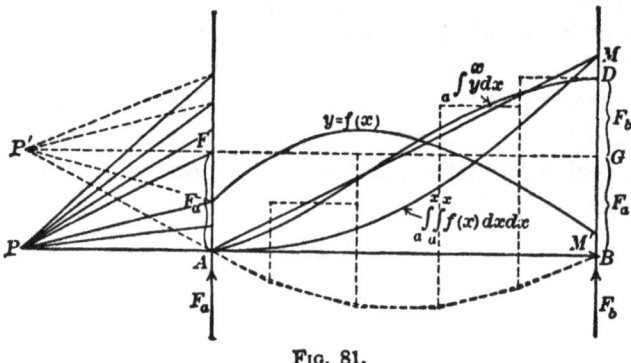

FIG. 81.

the sum of the two reactions a horizontal line through F will
divide BD into the two parts $BG = F_a$ and $GD = F_b$.

Shifting the position of P to P' on the horizontal line FG
and repeating the integration

$$\int_a^\xi Y dx,$$

we obtain a curve with equal ordinates at both ends. If we
begin at A it must end in B. Its ordinates are equal to the
difference between the ordinates of the chord AM and the curve
AM (Fig. 81), and represent the moment about any point of

the beam of all the forces on one side of the point (load and reaction).

The area of a closed curve may be found by integrating over the whole boundary. Suppose $x = a$ and $x = b$ to be the limits of the abscissas of the closed curve, the vertical $x = a$ touching the curve at A and the vertical $x = b$ at B (Fig. 82). By A and B the closed curve is cut in two, both parts connecting A and B. Let us denote the upper part by $y = f_1(x)$ and the lower part by $y = f_2(x)$. The whole area is then equal to the difference

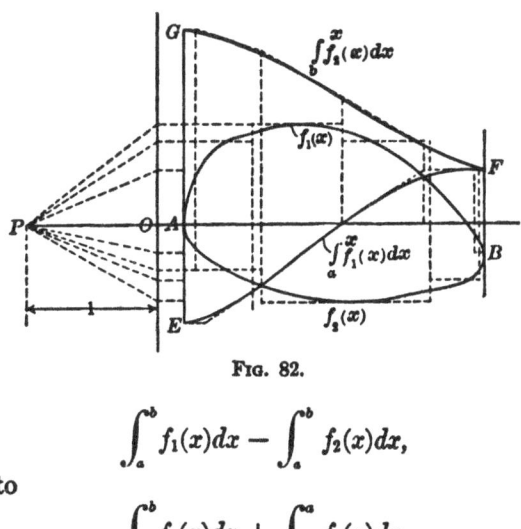

FIG. 82.

$$\int_a^b f_1(x)dx - \int_a^b f_2(x)dx,$$

or equal to

$$\int_a^b f_1(x)dx + \int_b^a f_2(x)dx.$$

We begin the integral curve over the upper part at the vertical $x = a$ at a point E, the ordinate of which is arbitrary, and draw the broken line as far as F on the vertical $x = b$ (Fig. 82). Then we integrate back again over the lower part, continuing the broken line from F to G. The line EG measured in the unit of length set down for the ordinates is equal to the area measured in units of area, this unit being a rectangle formed by PO and the unit of ordinates. That is to say, the area is equal to the area of a rectangle whose sides are PO and EG.

The method is not limited to the case drawn in Fig. 82, where the closed curve intersects any vertical not more than twice. A more complicated case is shown in Fig. 83. But in all those cases

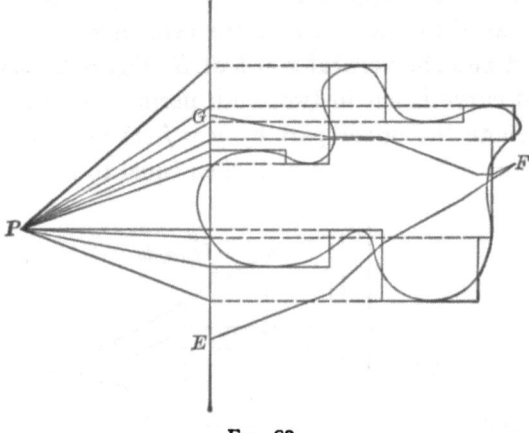

Fig. 83.

where the object is not to find the integral curve but only to find the value of the last ordinate the method, cannot claim to be of much use, because it cannot compete with the planimeter.

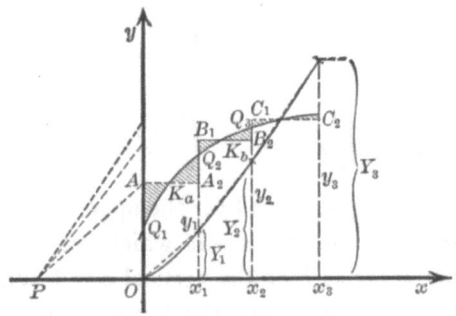

Fig. 84.

For the construction of the broken line we have drawn the steps in such a manner that the areas on both sides of the vertical part of a step between the curve and the stepping line are equal.

It would have also been admissible to construct the stepping line in such a way that the areas on both sides of the *horizontal* part of a step are equal (Fig. 84). Only the broken line would consist of a series of chords instead of a series of tangents of the integral curve. The points K_a, K_b, $\cdots$, where the horizontal parts of the steps intersect the curve would determine the abscissas of the points of the integral curve, where its direction is parallel to the direction of the broken line. But this forms very little help for drawing the integral curve. That is the reason why the former method where the broken line consists of a series of tangents is to be preferred. However where the object is only to find the last ordinate of the integral curve the two methods are equivalent.

§ 14. *Graphical Differentiation.*—The graphical differentiation of a function represented by a curve is not so satisfactory as the graphical integration because the values of the differential coefficient are generally not very well defined by the curve. The operation consists in drawing tangents to the given curve and drawing parallels through P to the tangents (Fig. 85). The points of intersection of these parallels with the axis of ordinates fur-

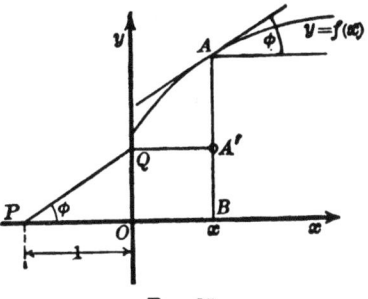

Fig. 85.

nish the ordinates of the curve representing the derivative. The abscissa to each ordinate coincides with the abscissa of the point of contact of the corresponding tangent. The principal difficulty is to draw the tangent correctly. As a rule it can be recommended to draw a tangent of a given direction and then mark its point of contact instead of trying to draw the tangent for a given point of contact. A method of finding the point of contact more accurately than by mere inspection consists in drawing a number of chords parallel to the tangent and to

bisect them. The points of bisection form a curve that intersects the given curve at the point of contact (Fig. 86). When a
number of tangents are drawn, their points of contact marked
and the points representing the differential coefficient constructed,

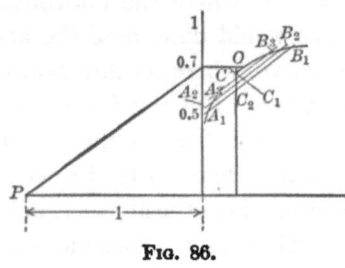

the derivative curve has to be
drawn through these points.
This may be done more accurately by means of the stepping
line. The horizontal parts of
the steps pass through the
points while the vertical parts
lie in the same vertical as the
point of intersection of two

Fig. 86.

consecutive tangents. The derivative curve connects the points
in such a way that the areas between it and the stepping line are
equal on both sides of the vertical parts of each step. Thus
the result of the graphical differentiation is exactly the same

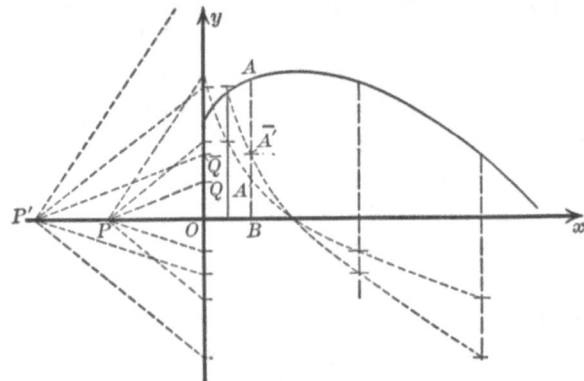

Fig. 87.

figure that we get by integration, only the operations are carried
out in the inverse order.

A change of the distance *PO* (Fig. 87) changes the ordinates
of the derivative curve in the same proportion and for the same
reason that it changes the ordinates of the integral curve when we

are integrating, but in the inverse ratio. Any change of the ordinate of P only shifts the curve up or down by an equal amount, so that if we at the same time change the axis of x and draw it through the new position of P the ordinates of the curve will remain the same and will represent the differential coefficient.

When a function $f(x, y)$ of two variables is given by a diagram showing the curves $f(x, y) =$ const. for equidistant values of $f(x, y)$ the partial differential coefficients can be found at any point x_0, y_0 by means of drawing curves whose ordinates represent $f(x, y_0)$ to the abscissa x or $f(x_0, y)$ to the abscissa y and applying the methods explained above. For this purpose a parallel is drawn to the axis of x, for instance, through the point x_0, y_0 and at the points where it intersects the curves $f(x, y) =$ const. ordinates are erected representing the values of $f(x, y_0)$ in any convenient scale. A smooth curve is then drawn though the points so found and the tangent of the curve at the point x_0 furnishes the differential coefficient $\partial f/\partial x$ for $x = x_0$, $y = y_0$.

The differential coefficients $\partial f/\partial x$, $\partial f/\partial y$ are best represented graphically by a straight line starting from the point x, y to which the differential coefficients correspond, and of such length and direction that its orthogonal projections on the axis of x and y are equal to $\partial f/\partial x$ and $\partial f/\partial y$. This line represents the gradient of the function $f(x, y)$ at the point x, y.[1] It is normal to the curve $f(x, y) =$ const. that passes through the point x, y, its direction being the direction of steepest ascent. Its length measures the slope of the surface $z = f(x, y)$ in the direction of steepest ascent. This is shown by considering the slope in any other direction. Let us change x and y by

$$r \cos \alpha, \quad r \sin \alpha$$

and consider the corresponding change

$$\Delta z = f(x + r \cos \alpha, y + r \sin \alpha) - f(x, y)$$

of the function. By Taylor's theorem we can write it

[1] See Chap. II, § 10.

$$\frac{\partial f}{\partial x}\, r \cos \alpha + \frac{\partial f}{\partial y}\, r \sin \alpha + \text{terms of higher order in } r,$$

α is the direction from the point x, y to the new point $x + r \cos \alpha$, $y + r \sin \alpha$ and r is the distance of the two points. Dividing Δz by r and letting r approach to zero we find

$$\lim \frac{\Delta z}{r} = \frac{\partial f}{\partial x} \cos \alpha + \frac{\partial f}{\partial y} \sin \alpha.$$

This expression measures the slope of the surface $z = f(xy)$ in the direction α. Now let us introduce the length l and the angle λ of the gradient, and write

$$\frac{\partial f}{\partial x} = l \cos \lambda, \quad \frac{\partial f}{\partial y} = l \sin \lambda.$$

Then we have

$$\frac{\partial f}{\partial x} \cos \alpha + \frac{\partial f}{\partial y} \sin \alpha = l \cos (\alpha - \lambda).$$

That is to say, the slope in any direction α is proportional to $\cos (\alpha - \lambda)$, it is a maximum in the direction of the gradient ($\alpha = \lambda$) and zero in a direction perpendicular to it and negative in all directions that form an obtuse angle with it. When all three coördinates are measured in the same unit, the length of l measured in this unit is equal to the tangent of the angle of steepest ascent. Hence the length of the gradient varies with the unit of length. When the unit of length in which the values of $f(xy)$ are plotted is kept unaltered, while we change the unit of length corresponding to the values x and y, the length of the gradient varies with the square of the unit of length.

§ 15. *Differential Equations of the First Order.*—In the problem of solving a differential equation of the first order

$$\frac{dy}{dx} = f(x, y)$$

by graphical methods the first question is how to represent the differential equation graphically. If x and y are meant to be the values of rectangular coördinates, the geometrical meaning of the differential equation is that at every point x, y, where $f(x, y)$ is defined, the equation prescribes a certain direction for the curve that satisfies it. Let us suppose curves drawn through all those points for which $f(x, y)$ has certain constant values. Each curve then corresponds to a certain direction or the opposite direction. Let us distinguish the curves by different numbers or letters and let us draw a pencil of rays together with the curves and mark the rays with the same numbers or letters in such a way that each of them shows the direction corresponding to the

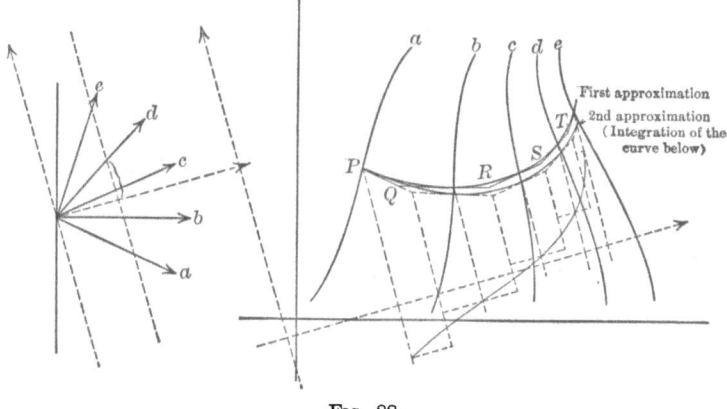

FIG. 88.

curve marked with that particular number or letter (Fig. 88). Our drawing of course only comprises a certain region in which we propose to find the curves satisfying the differential equation. It may be that $f(xy)$ is defined beyond the boundaries of our drawing. Those regions have to be dealt with separately.

The graphical representation of the differential equation in the region considered consists in the correspondence between the curves and the rays. It is important to observe that this representation is independent of the system of coördinates by means of which we have deduced the curves from the equation

$$\frac{dy}{dx} = f(xy).$$

We can now introduce any system of coördinates ξ, η and find from our drawing the equation

$$\frac{d\eta}{d\xi} = \varphi(\xi\eta),$$

that is to say, we can find the value of $\varphi(\xi, \eta)$ at any point ξ, η of our drawing. If, for instance, the unit of length is the same for ξ and η we draw a line through the center of the pencil of rays in the direction of the positive axis of ξ and a line perpendicular to it at the distance 1 from the center. The segment on the second line between the first line and the point of intersection with one of the rays measured in units of length and counted positive in the direction of positive η furnishes the value of $\varphi(\xi, \eta)$ for all the points ξ, η corresponding to that particular ray. In this respect the graphical representation of a differential equation is superior to the analytical form, in which certain coördinates are used and the transformation to another system of coördinates requires a certain amount of calculation.

Now let us try to find the curve through a given point P on the curve marked (a) (Fig. 88) that satisfies the differential equation. We begin by drawing a series of tangents of a curve that is meant to be a first approximation. Through P we draw a parallel to the ray (a) as far as the point Q somewhere in the middle between the curves (a) and (b). Through Q we draw a parallel to the ray (b) as far as R somewhere in the middle between the curves (b) and (c). Through R we again draw a parallel to the ray (c) and so on. The curve touching this broken line at the points of intersection with the curves (a), (b), $\cdots$ is a first approximation. But we need not draw this curve. In order to find a better approximation we introduce a rectangular system of coördinates x, y, laying the axis of x somewhat in the mean direction of the broken line. Let us denote by y_1 the function of x that corresponds to the curve forming the first approximation. The second approximation y_2 is then obtained as an integral curve of $f(x, y_1)$, that is, of dy_1/dx

$$y_2 = y_p + \int_{x_p}^{x} f(x, y_1) dx,$$

denoting by x_p, y_p, the coördinates of P. For this purpose the curve whose ordinates are equal to $f(x, y_1)$ or dy_1/dx has to be constructed first. The values of $f(x, y_1)$ are found immediately at the points where the first approximation intersects the curve (a), (b) $\cdots$ by differentiation in the way described above. A line is drawn through the center of the pencil of rays parallel to the axis of x and a line perpendicular to it at a convenient distance from the center. This distance is chosen as the unit of length. The points of intersection of this line with the rays determine segments whose lengths are equal to the values of $f(x, y_1)$ on the corresponding curves. These values are plotted as ordinates to the abscissas of the points where the first approximation intersects the curves (a), (b), $\cdots$ and a curve

$$Y = f(x, y_1)$$

is drawn (Fig. 88). This curve is integrated graphically beginning at the point P and the integral curve is a second approximation. Again we need not draw the curve. The broken line suffices, if we intend to construct a third approximation. In this case we have to repeat the foregoing operation. This can now be performed much quicker than in the first case because the values of $f(x, y)$ on the curves (a), (b), $\cdots$ have already been constructed and are at our disposal. In order to find the curve

$$Y = f(x, y_2)$$

we have only to shift the same ordinates to new abscissas and make these coincide with the abscissas of the points where the second approximation intersects the curves (a), (b), $\cdots$. The curve

$$Y = f(x, y_2)$$

is then drawn and integrated graphically, beginning at the point P.

Suppose now the integral curve did not differ from the second approximation, it would mean that

$$y_2 = y_p + \int_{x_p}^{x} f(x, y_2)dx,$$

or that

$$\frac{dy_2}{dx} = f(x, y_2),$$

that is to say, that y_2 satisfies the differential equation.

If there is a perceptible difference the integral curve represents a third approximation. It has been shown by Picard that proceeding in this way we find the approximations (under a certain condition to be discussed presently) converging to the true solution of the differential equation, so that after a certain number of operations the error of the approximation must become imperceptible.

Denoting by y_n the function of the nth approximation we have

$$y_{n+1} = y_p + \int_{x_n}^{x} f(x, y_n)dx.$$

The true solution with the same initial conditions $y = y_p$ for $x = x_p$ satisfies the equation

$$y = y_p + \int_{x_p}^{x} f(x, y)dx.$$

Hence

$$y_{n+1} - y = \int_{x_p}^{x} [f(x, y_n) - f(x, y)]dx,$$

or

$$y_{n+1} - y = \int_{x_p}^{x} \frac{f(x, y_n) - f(x, y)}{y_n - y} (y_n - y)dx.$$

Let us now suppose that the absolute value of

$$\frac{f(x, y_n) - f(x, y)}{y_n - y},$$

for all the values of x, y, y_n within the considered region does not surpass a certain limit M, then it follows that a certain relation must exist between the maximum error of y_n, which we denote by e_n and the maximum error of y_{n+1}, which we denote by e_{n+1}. The absolute value of the integral not being larger than

$$Me_n \,|\, x - x_n \,|$$

($|\, x - x_n \,|$ denoting the absolute value of $x - x_n$) we have

$$e_{n+1} \leqq M \,|\, x - x_n \,|\, e_n.$$

Hence as long as the distance $x - x_n$ over which the integration is performed is so small that

$$M \,|\, x - x_n \,| \leqq k < 1,$$

k being a constant smaller than one, the error of y_{n+1} cannot be larger than a certain fraction of the maximum error of y_n. But in the same way it follows that the error of y_n cannot be larger than the same fraction of the maximum error of y_{n-1}, and so on, so that

$$e_{n+1} \leqq ke_n \leqq k^2 e_{n-1} \cdots \leqq k^n e_1.$$

But as e_1 is a constant and k a constant smaller than one, $k^n e_1$ must be as small as we please for a sufficient large value of n. That is to say, the approximations converge to the true solution.

M being a given constant the condition of convergence

$$M \,|\, x - x_p \,| \leqq k < 1$$

limits the extent of our integration in the direction of the axis of x. But it does not limit our progress. From any point P' that we have reached with sufficient accuracy we can make a fresh start, choosing a new axis of x suited to the new situation. As a rule it does not pay to trouble about the value of M and to try to find the extent of the convergence by the help of this value. The actual construction of the approximations will show clearly enough how far to extend the integration. As far as two consecutive approximations show no difference they represent the true curve.

Suppose that

$$\frac{f(x, y_n) - f(x, y)}{y_n - y}$$

has the same sign for all values x, y, y_n concerned. Say it is negative. Suppose further that $y_n - y$ is of the same sign for the whole extent of the integration

$$y_{n+1} - y = \int_{x_p}^{x} \frac{f(x, y_n) - f(x, y)}{y_n - y} (y_n - y)dx;$$

that is to say, the approximative curve y_n is all on one side of the true curve. Then if $x - x_p$ is positive, $y_{n+1} - y$ must evidently be of the opposite sign from $y_n - y$, or the approximative curve y_{n+1} is all on the other side of the true curve from y_n. For these and all following approximations the true curve must lie between two consecutive approximations. If the first approximation y_1 is all on one side of the true curve the theorem holds for any two consecutive approximations. This is very convenient for the estimation of the error.

In Fig. 88

$$\frac{f(x, y_n) - f(x, y)}{y_n - y}$$

is negative from the point P as far as somewhere near S. The first approximation is all on the upper side of the true curve. Therefore the second approximation must be below the true curve at least as far as somewhere near S.

When the sign is positive the same theorem holds for negative values of $x - x_p$. If the integration has been performed in the positive direction of x, it may be a good plan to check the result by integrating backwards, starting from a point that has been reached and to try if the curve gets back to the first starting point. In this direction we profit from the advantage of the true curve lying between consecutive approximations and are better able to estimate the accuracy of our drawing.

We have seen that the convergence depends on the maximum

absolute value of

$$\frac{f(x, y_n) - f(x, y)}{y_n - y}$$

for all values of x, y, y_n concerned. In order to find the maximum value we may as well onsider

$$\frac{\partial f}{\partial y}$$

for all values of x, y within the region considered. For if we assume $\partial f/\partial y$ to be a continuous function of y, it follows that the quotient of differeces

$$\frac{f(x, y_n) - f(x, y)}{y_n - y}$$

must be equal to $\partial f/\partial y$ taken for the same value of x and a value of y between y and y_n. This is immediately seen by plotting $f(x, y)$ as ordinate to the abscissa y for a fixed value of x. The value of the quotient of differences is determined by the slope of the chord between the two points of abscissas y and y_n. The slope of the chord is equal to the slope of the curve at a certain point between the end of the chord. The value of $\partial f/\partial y$ at this point is equal to the vlue of

$$\frac{f(x, y_n) - f(x, y)}{y_n - y}.$$

Now let us consider how the coördinate system may be chosen in order to make $\partial f/\partial y$ as small as possible and thus obtain the best convergence. Fo this purpose let us investigate how the value of $\partial f/\partial y$ changes at a certain point, when the system of coördinates is changed.

Let us start with a given system of rectangular coördinates ξ, η with which the differntial equation is written

$$\frac{d\eta}{d\xi} = \varphi(\xi, \eta).$$

The direction of the curve satisfying the differential equation forms a certain angle α with the positive axis of ξ determined by

$$\operatorname{tg} \alpha = \frac{d\eta}{d\xi} = \varphi(\xi, \eta)$$

(assuming the coördinates to be measured in the same unit). Now let us introduce a new system of rectangular coördinates x, y connected with the system ξ, η by the equations

$$x = \xi \cos \omega + \eta \sin \omega,$$
$$y = -\xi \sin \omega + \eta \cos \omega,$$

which are equivalent to

$$\xi = x \cos \omega - y \sin \omega,$$
$$\eta = x \sin \omega + y \cos \omega,$$

ω being the angle between the positive direction of x and the positive direction of ξ, counted from ξ towards x in the usual way.

The angle formed by the direction of the curve with the positive direction of the axis of x is $\alpha - \omega$, and therefore

$$\frac{dy}{dx} = \operatorname{tg}(\alpha - \omega) = f(x, y).$$

Consequently we obtain for a given value of ω

$$\frac{\partial f}{\partial y} = \frac{1}{\cos^2(\alpha - \omega)} \cdot \frac{\partial \alpha}{\partial y},$$

or remembering that α is given as a function of ξ and η,

$$\frac{\partial f}{\partial y} = \frac{1}{\cos^2(\alpha - \omega)} \cdot \left(-\frac{\partial \alpha}{\partial \xi} \sin \omega + \frac{\partial \alpha}{\partial \eta} \cos \omega \right).$$

For simplicity's sake we shall assume that the axis of ξ is the tangent of the curve $\varphi(\xi, \eta) = $ const. that passes through the given point, so that $\partial \alpha / \partial \xi = 0$.

We then have

$$\frac{\partial f}{\partial y} = \frac{1}{\cos^2(\alpha - \omega)} \frac{\partial \alpha}{\partial \eta} \cos \omega,$$

and our object is to find how $\partial f / \partial y$ varies for different values of

ω. The value of $\partial\alpha/\partial\eta$ is independent of ω; it denotes the value of the gradient of α, which we represent by a straight line drawn from the origin A (Fig. 89) perpendicular to the curve $\alpha = $ const. or $\varphi(\xi, \eta) = $ const.

It is no restriction to assume the value of $\partial\alpha/\partial\eta$ positive; it only means that the direction of the positive axis of η is chosen

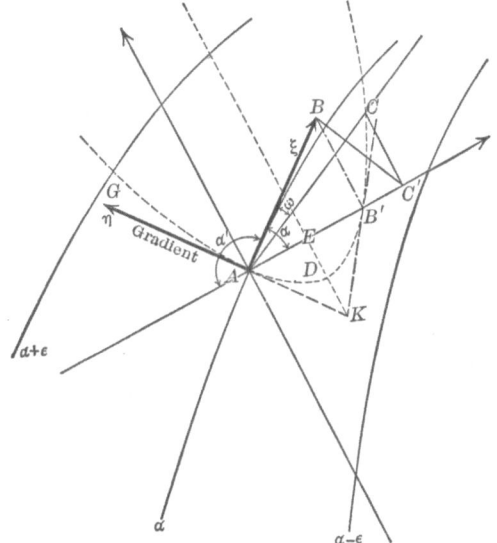

FIG. 89.

in the direction of the gradient. Let us draw the line AB (Fig. 89) in the direction of the positive axis of ξ and of the same length as the gradient.

In order to show the values of $\partial f/\partial y$ for the different positions of the axis of x let us lay off the value of $\partial f/\partial y$ as an abscissa. For instance for $\omega = \alpha$, $\partial f/\partial y$ assumes the value

$$\frac{\partial\alpha}{\partial\eta}\cos\alpha.$$

The abscissa corresponding to this value is AB' (Fig. 89), the

10

orthogonal projection of AB on the axis of x. For any other
position AC (Fig. 89) corresponding to some other value of ω,
we find $\partial\alpha/\partial\eta \cos \omega$ by orthogonal projection of AB on AC. Then
the division by $\cos (\alpha - \omega)$ furnishes AC' and a second division
by $\cos (\alpha - \omega)$ leads to AC. Thus a certain curve can be
constructed whose polar coördinates are $r = \partial f/\partial y$ and ω, the
equation in polar coördinates being

$$r = \frac{\partial\alpha}{\partial\eta} \cdot \frac{\cos \omega}{\cos^2 (\alpha - \omega)} \quad \text{or} \quad [r \cos (\alpha - \omega)]^2 = \frac{\partial\alpha}{\partial\eta} r \cos \omega.$$

In rectangular coördinates ξ, η the equation assumes the form

$$(\cos \alpha \xi + \sin \alpha \eta)^2 = \frac{\partial\alpha}{\partial\eta} \xi.$$

This shows that the equation is a parabola, the axis of which is
perpendicular to the direction α. AB' is a chord and the gradient

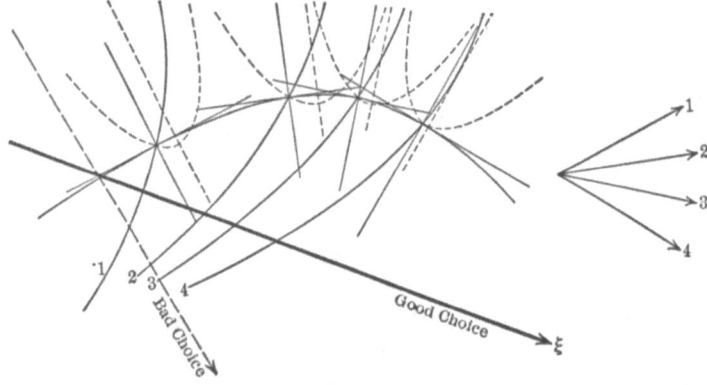

FIG. 90.

AG is a tangent of the parabola.   Bisecting AB' in E, drawing EK
perpendicular to AB' as far as the axis of η and bisecting EK in
D, we find D the apex of the parabola.   The three points A, B', D
together with the gradient will suffice to give us an idea of the
size and sign of $\partial f/\partial y$ for the different positions of the positive
axis of x.

$\partial f/\partial y$ vanishes when the axis of x is perpendicular to the curve $\alpha =$ const., so that it seems as if this were the most favorable position. We must, however, bear in mind that the axis of x is kept unaltered for a certain interval of integration. When we pass on to other points the axis of x is no longer perpendicular to the curve $\alpha =$ const. there. The position of the axis of x is good when the average value of $\partial f/\partial y$ is small. In Fig. 90 the parabolas are constructed for a number of points on the first approximation of a curve satisfying the differential equation.

If we want to make use of the parabolas to give us the numerical values of $\partial f/\partial y$ the unit of length must also be marked in which the coördinates are measured. The numerical value of $\partial f/\partial y$ varies as the unit of length and therefore the length of the line representing it must vary as the square of the unit of length. But if we draw a line whose length measured in the same unit is equal to $\dfrac{1}{\partial f/\partial y}$, this line would be independent of the unit of length. For if l is the line representing the unit of length and l', l'' the lines representing the values $\partial f/\partial y$ and $\dfrac{1}{\partial f/\partial y}$, $\partial f/\partial y$ would be the ratio l'/l and $\dfrac{1}{\partial f/\partial y}$ the ratio l''/l; hence $l'' = l^2/l'$. Since l' varies as l^2 with the change of the unit of length l'' is independent of the unit of length. This line l'' represents the limit beyond which the product

$$\frac{\partial f}{\partial y} \cdot l''$$

becomes greater than 1. If $\partial f/\partial y$ remained the same this would mean the limit beyond which the convergence of the process of approximation ceases. We might lay off the length of $\dfrac{1}{\partial f/\partial y}$ in the different directions in the same way as $\partial f/\partial y$ has been laid off. The result is a curve corresponding, point by point, to the parabola, the image of the parabola according to the relation of reciprocal radii. But all these preparations as a rule would not pay. It is better to attack the integration at once with an axis of x somewhat perpendicular to the curves α = const. as long as the direction of the curve forms a considerable angle with the curve α = const. and to lose no time in troubling about the very best position. The convergence will show itself, when the operations are carried out. When the angle between the direction of the curve that satisfies the differential equation and the curve α = const. becomes small the apex of the parabola moves far away and when the direction coincides with that of the curve α = const. the parabola degenerates into two parallel lines perpendicular to the direction of the curve α = const. In this case the best position for the axis of x is in the direction of the curve α = const. Without going into any detailed investigation about the best position of the axis of x we can establish the general rule not to make the axis of x perpendicular to the direction of the curve satisfying the differential equation, that is to say, not to make it parallel to the axis of the parabola. But we hardly need pronounce this rule. In practice it would enforce its own observance, because for that position of the axis of x not only $\partial f/\partial y$ but also $f(x, y)$ are infinite and it would become impossible to plot the curve $Y = f(x, y_1)$.

There is another graphical method of integrating a differential equation of the first order

$$\frac{dy}{dx} = f(x, y),$$

which in some cases may well compete with the first method. Like the first it is the analogue of a certain numerical method.

The numerical method starts from given values x, y and calculates the change of y corresponding to a certain small change of x. Let h be the change of x and k the change of y, so that $x + h$, $y + k$ are the coördinates of a point on the curve satisfying the differential equation and passing through the point x, y. k is calculated in the following manner. We calculate in succession four values k_1, k_2, k_3, k_4 by the following equations—

$$k_1 = f(x, y)h,$$

$$k_2 = f\left(x + \frac{h}{2}, y + \frac{k_1}{2}\right)h,$$

$$k_3 = f\left(x + \frac{h}{2}, y + \frac{k_2}{2}\right)h,$$

$$k_4 = f(x + h, y + k_3)h.$$

We then form the arithmetical means

$$p = \frac{k_2 + k_3}{2} \quad \text{and} \quad q = \frac{k_1 + k_4}{2},$$

and find with a high degree of approximation as long as h is not too large

$$k = p + \tfrac{1}{3}(q - p).[1]$$

The new values

$$X = x + h, \quad Y = y + k$$

are then substituted for x and y and in the same way the coördinates of a third point are calculated and so on.

This calculation may be performed graphically in a profitable manner, if the function $f(x, y)$ is represented in a way suited to

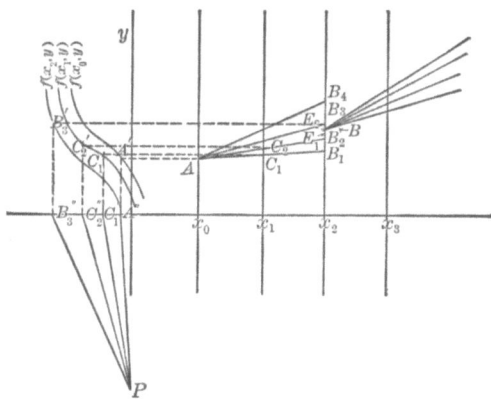

FIG. 91.

[1] See W. Kutta, *Zeitschrift für Mathematik und Physik*, Vol. 46, p. 443.

the purpose. Let us suppose a number of equidistant parallels to the axis of ordinates: $x = x_0$, $x = x_1$, $x = x_2$, $x = x_3$, $\cdots$. Along these lines $f(x, y)$ is a function of y. Let us lay off the values of $f(x, y)$ as ordinates to the abscissa y, the axis of y being taken as the axis of abscissas. We thus obtain a number of curves representing the functions $f(x_0, y)$, $f(x_1, y)$, $f(x_2, y)$, $\cdots$. Starting from a point $A(x_0, y_0)$ on the first vertical $x = x_0$ (Fig. 91) we proceed to a point B_1 on the vertical $x = x_2$ in the following way. By drawing a horizontal line through A we find the point A' on the curve representing $f(x_0, y)$. Its ordinate is equal to $f(x_0, y_0)$. Projecting the point A' onto the axis of x we find A'' and draw the line PA''. P is a point on the negative side of the y-axis and PO is equal to the unit of length by which the lines representing $f(x, y)$ are measured. Thus

$$OA''/PO = f(x_0, y_0).$$

Now we draw AB_1 perpendicular to PA'', so that if h and k_1 denote the differences of the coördinates of A and B, we have

$$k_1/h = OA''/PO,$$

$$k_1 = f(x_0, y_0)h.$$

From C_1 the point of intersection of the line AB_1 and the vertical $x = x_1$ we find C_1' and C_1'' in the same way as we found A' and A'' from A, only that C_1' is taken in the curve representing the values of $f(x_1, y)$, and draw the line AB_2 perpendicular to PC_1''. Denoting the difference of the ordinates of A and B_2 by k_2 we have

$$\frac{k_2}{h} = \frac{OC_1''}{PO} = f\left(x_0 + \frac{h}{2}, y_0 + \frac{k_1}{2}\right),$$

or

$$k_2 = f\left(x_0 + \frac{h}{2}, y_0 + \frac{k_1}{2}\right)h.$$

From C_2 the point of intersection of the line AB_2 and the vertical $x = x_1$ we find in the same way a point B_3 on the vertical

$x = x_2$ and the difference k_3 between the ordinate of B_3 and that of A is

$$k_3 = f\left(x_0 + \frac{h}{2}, y_0 + \frac{k_2}{2}\right)h.$$

From B_3 we pass horizontally to B_3' on the curve representing $f(x_2 y)$ and vertically down to B_3''. The line AB_4 is then drawn perpendicular to PB_3'', so that the difference k_4 between the ordinates of B_4 and A is

$$k_4 = f(x_0 + h, y + k_3)h.$$

The bisection of $B_2 B_3$ and of $B_1 B_4$ gives us the points E_1 and E_2 and the point B is taken between E_1 and E_2, so that its distance from E_1 is half its distance from E_2. The point B is with a high degree of approximation a point of the curve that passes through A and satisfies the differential equation.

B is then taken as a new point of departure instead of A, and in this manner a series of points of the curve are found.

In order to get an idea of the accuracy attained the distance of the vertical lines is altered. For instance, we may leave out the verticals $x = x_1$ and $x = x_3$, and reach the point on the vertical $x = x_4$ in one step instead of two. The error of this point should then be about sixteen times as large as the error on the same vertical reached by two steps, so that the error of the latter should be about one-fifteenth of the distance of the two. If their distance is not appreciable the smaller steps are evidently unnecessarily small.

The values of $f(x, y)$ may become so large that an inconveniently small unit of length must be applied to plot them. In this case x and y have to change parts and the differential equation is written in the form

$$\frac{dx}{dy} = \frac{1}{f(x, y)}.$$

The values of $1/f(x, y)$ are then plotted for equidistant values of

y as ordinates to the abscissa x and the constructions are changed accordingly.

§ 16. *Differential Equations of the Second and Higher Orders.*— Differential equations of the second order may be written in the form

$$\frac{d^2y}{dx^2} = f\left(x, y, \frac{dy}{dx}\right).$$

Let us introduce the radius of curvature instead of the second differential coefficient. Suppose we pass along a curve that satisfies the equation and the direction of our motion is determined by the angle α it forms with the positive axis of x (counted in the usual way from the positive axis of x through ninety degrees to the positive axis of y and so on), s being the length of the curve counted from a certain point from which we start. We then have

$$\frac{dy}{dx} = \operatorname{tg} \alpha, \quad \frac{dx}{ds} = \cos \alpha.$$

Consequently

$$\frac{d^2y}{dx^2} = \frac{1}{\cos^2 \alpha} \cdot \frac{d\alpha}{dx} = \frac{1}{\cos^3 \alpha} \cdot \frac{d\alpha}{ds},$$

or

$$\frac{d\alpha}{ds} = \cos^3 \alpha \frac{d^2y}{dx^2}.$$

$d\alpha/ds$ measures the "curvature," the rate of change of direction as we pass along the curve, counted positive when the change takes place to the side of greater values of α (if the positive axis of x is drawn to the right and the positive axis of y upwards a positive value of $d\alpha/ds$ means that the path turns to the left). Let us count the radius of curvature with the same sign as $d\alpha/ds$ and let us denote it by ρ. Then we have

$$\frac{1}{\rho} = \cos^3 \alpha f(x, y, \operatorname{tg} \alpha).$$

Thus the differential equation of the second order may be said to give the radius of curvature as a function of x, y, α, that is to say, as a function of place and direction.

Let us assume that this function of three variables is represented by a diagram, so that the length and sign of ρ may quickly be obtained for any point and any direction.

Starting from any given point in any given direction we can then approximate the curve satisfying the differential equation by a series of circular arcs. Let A (Fig. 92) be the starting point. We make $M_a A$ perpendicular to the given direction and equal to ρ in length. For positive values of ρ, M_a must be on the positive side of the given direction, for negative

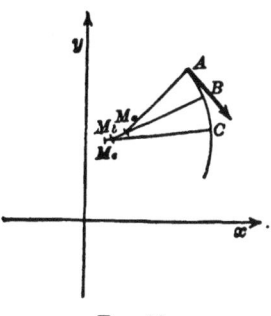

FIG. 92.

values on the negative side. M_a is the center of curvature for the curve at A. With M_a as center and $M_a A$ as radius we draw a circular arc AB and draw the line BM_a. On this line or on its production we mark the point M_b at a distance from B equal to the value of ρ that corresponds to B and to the direction in which the circular arc reaches B. With M_b as center and $M_b B$ as radius we draw a circular arc BC and so on.

The true curve changes its radius of curvature continuously, while our approximation changes it abruptly at the points $A, B, C, \cdots$. The smaller the circular arcs the less will accurately-drawn circular arcs deviate from the curve. But it must be kept in mind that small errors cannot be avoided, when passing from one arc to the next. Hence, if the arcs are taken very small so that their number for a given length of curve increases unduly, the accuracy will not be greater than with somewhat longer arcs. The best length cannot well be defined mathematically; it must be left to the experience of the draughtsman.

Some advantage may be gained by letting the centers and the radii of the circular arcs deviate from the stated values. The circular arc AB (Fig. 92) is evidently drawn with too small a radius because the radius of the curve increases towards B. If we had taken the radius equal to M_bB it would have been too large. A better approximation is evidently obtained by making the radius of the first circular arc equal to the mean of M_aA and M_bB, and the direction with which it reaches B will also be closer to the right direction.

To facilitate the plotting an instrument may be used consisting of a flat ruler with a hole on one end for a pencil or a capillary tube or any other device for tracing a line. A straight line with a scale is marked along the middle of the ruler and a little tripod of sewing needles is placed with one foot on the line and two feet on the paper. Thus the pencil traces a circular arc. When the radius is changed, the ruler is held in its position by pressing it against the paper until the tripod is moved to a new position. By this device the pencil must continue its path in exactly the same direction, while with the use of ordinary compasses it is not easy to avoid a slight break in the curve at the joint of two circular arcs.

Another method consists in a generalization of the method for the graphical solution of a differential equation of the first order.

A differential equation of the second order

$$\frac{d^2y}{dx^2} = g\left(x, y, \frac{dy}{dx} \right)$$

may be written in the form of two simultaneous equations of the first order:

$$\frac{dy}{dx} = z,$$

$$\frac{dz}{dx} = g(x, y, z).$$

Let us consider the more general form, in which the differential coefficients of two functions y, z of x are given as functions of x, y, z:

$$\frac{dy}{dx} = f(x, y, z),$$

$$\frac{dz}{dx} = g(x, y, z).$$

We may interpret x, y, z as the coördinates of a point in space and the differential equation as a law establishing a certain direction or the opposite at every point in space where $f(x, y, z)$ and $g(x, y, z)$ are defined. A curve in space satisfies the differential equation, when it never deviates from the prescribed direction. Its projection in the xy plane represents the function y and its projection in the xz plane represents the function z.

Let us represent y and z as ordinates and x as abscissa in the same plane with the same system of coördinates. Any point in

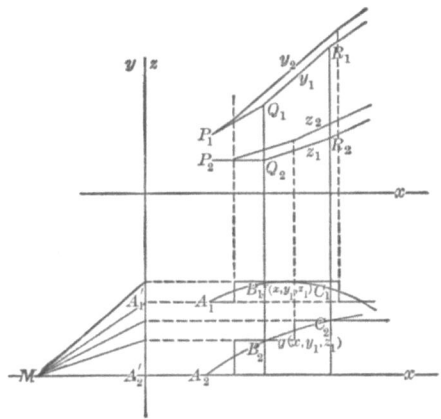

Fig. 93.

space is represented by two points with the same abscissa. The functions $f(x, y, z)$ and $g(x, y, z)$ we suppose to be given either by diagrams or by certain methods of construction or calculation. For any point that we have to deal with, the values of $f(x, y, z)$ and $g(x, y, z)$ are plotted as ordinates to the abscissa x, but for clearness sake not in the same system of coördinates as y and z, but in another system with the same axis of ordinates and an axis of x parallel to the first and removed far enough so that the drawings in the two systems do not interfere with one another.

Starting from a certain point $P(x_p, y_p, z_p)$ in space we represent it by the two points $P_1(x_p, y_p)$ and $P_2(x_p, z_p)$ in the first system and the values of $f(x_p, y_p, z_p)$ and $g(x_p, y_p, z_p)$ by the two points A_1 and A_2 in the second system of coördinates (Fig. 93). The points A_1 and A_2 determine certain directions MA_1', and MA_2' of the curves x, y and x, z, the point M (Fig. 93) being placed at a distance from the axis of ordinates equal to the unit of length by which the ordinates representing $f(x, y, z)$ and $g(x, y, z)$ are measured. Through P_1 and P_2 we draw parallels to MA_1' and MA_2' as far as Q_1 and Q_2 with the coördinates x_q, y_q and x_q, z_q. With these coördinates the values $f(x_q, y_q, z_q)$ and $g(x_q, y_q, z_q)$ are determined, which we represent by the ordinates of the points B_1, B_2. These points again determine certain directions parallel to which the lines Q_1R_1 and Q_2R_2 are drawn, etc. In this manner we find first approximations y_1 and z_1 for the functions y and z and corresponding to these approximations we find curves representing $f(x, y_1, z_1)$ and $g(x, y_1, z_1)$. These curves are now integrated graphically, the integral curve of $f(x, y_1, z_1)$ beginning at P_1 and the integral curve of $g(x, y_1, z_1)$ at P_2 and lead to second approximations y_2 and z_2:

$$y_2 = y_p + \int_{x_p}^{x} f(x, y_1, z_1)dx,$$

$$z_2 = z_p + \int_{x_p}^{x} g(x, y_1, z_1)dx.$$

For these second approximations the values of $f(x, y_2, z_2)$ and $g(x, y_2, z_2)$ are determined at a number of points along the curves x, y_2 and x, z_2 sufficiently close to construct the curves representing $f(x, y_2, z_2)$ and $g(x, y_2, z_2)$. By their integration a third approximation y_3, z_3 is obtained

$$y_3 = y_p + \int_{x_p}^{x} f(x, y_2, z_2)dx,$$

$$z_3 = z_p + \int_{x_p}^{x} g(x, y_2, z_2)dx,$$

and so on as long as a deviation of an approximation from the one before can still be detected. As soon as there is no deviation for a certain distance $x - x_p$ the curve represents the true solution (as far as the accuracy of the drawing goes). The curve is continued by taking its last point as a new starting point for a similar operation.

The distance over which the integral is taken can in general not surpass a certain limit where the convergence of the approximations ceases. But we are free to make it as small as we please and accordingly increase the number of operations to reach a given distance. It is evidently not economical to make it too small. On the contrary, we shall choose it as large as possible without unduly increasing the number of approximations.

In the case of a differential equation

$$\frac{d^2y}{dx^2} = g\left(x, y, \frac{dy}{dx}\right)$$

we have $f(x, y, z) = z$, and the curve z, x is identical with the curve representing the values of $f(x, y, z)$. We shall therefore draw it only once.

The proof of the convergence of the approximations is almost the same as in the case of the differential equation of the first order.

For the $n + 1^{st}$ approximation we have

$$y_{n+1} = y_p + \int_{x_p}^{x} f(x, y_n, z_n)dx; \quad z_{n+1} = z_p + \int_{x_p}^{x} g(x, y_n, z_n)dx.$$

For the true curve that passes through the point x_p, y_p, z_p we find by integration

$$y = y_p + \int_{x_p}^{x} f(x, y, z)dx; \quad z = z_p + \int_{x_p}^{x} g(x, y, z)dx;$$

hence

$$y_{n+1} - y = \int_{x_p}^{x} [f(x, y_n, z_n) - f(x, y, z)]dx;$$

$$z_{n+1} - z = \int_{x_p}^{x} [g(x, y_n, z_n) - g(x, y, z)]dx.$$

Now let us write

$$f(x, y_n, z_n) - f(x, y, z) = \frac{f(x, y_n, z_n) - f(x, y, z_n)}{y_n - y} (y_n - y)$$
$$+ \frac{f(x, y, z_n) - f(x, y, z)}{z_n - z} (z_n - z),$$

and similarly

$$g(x, y_n, z_n) - g(x, y, z) = \frac{g(x, y_n, z_n) - g(x, y, z_n)}{y_n - y} (y_n - y)$$
$$+ \frac{g(x, y, z_n) - g(x, y, z)}{z_n - z} (z_n - z).$$

The quotients of differences

$$\frac{f(x, y_n, z_n) - f(x, y, z_n)}{y_n - y}$$

and the three others are equal to certain values of $\partial f/\partial y$, $\partial f/\partial z$, $\partial g/\partial y$, $\partial g/\partial z$ for values of y, z between y and y_n and between z and z_n (y, y_n, z, z_n not excluded). Let us assume that for the region of all the values of x, y, z concerned the absolute value of $\partial f/\partial y$ and $\partial f/\partial z$, is not greater than M_1, and that of $\partial g/\partial y$ and $\partial g/\partial z$ not greater than M_2, and that δ_n, ϵ_n denote the maximum of the absolute values of $y - y_n$ and $z - z_n$ in the interval x_p to x. Then it follows that the absolute values of

$$f(x, y_n, z_n) - f(x, y, z) \quad \text{and} \quad g(x, y_n, z_n) - g(x, y, z)$$

are not greater than

$$M_1(\delta_n + \epsilon_n) \quad \text{and} \quad M_2(\delta_n + \epsilon_n).$$

Hence for the maximum values of $y_{n+1} - y$ and $z_{n+1} - z$, which are denoted by δ_{n+1} and ϵ_{n+1} we obtain the limits

$$\delta_{n+1} \leqq M_1(\delta_n + \epsilon_n) \, | x - x_p |, \quad \epsilon_{n+1} \leqq M_2(\delta_n + \epsilon_n) \, | x - x_p |,$$

and

$$\delta_{n+1} + \epsilon_{n+1} \leqq (M_1 + M_2) \, | x - x_p | \, (\delta_n + \epsilon_n).$$

If therefore the interval $x - x_p$ of the integration is so far reduced that

$$(M_1 + M_2) \, | x - x_p | \leqq k < 1,$$

$\delta_{n+1} + \epsilon_{n+1}$ is not larger than the fraction k of $(\delta_n + \epsilon_n)$, but from the same reason

$$(\delta_n + \epsilon_n) \leq k(\delta_{n-1} + \epsilon_{n-1}), \quad (\delta_{n-1} + \epsilon_{n-1}) < k(\delta_{n-2} + \epsilon_{n-2}), \text{ etc.;}$$

therefore

$$\delta_{n+1} + \epsilon_{n+1} \leq k^n(\delta_1 + \epsilon_1).$$

That is to say, for a sufficiently large value of n δ_{n+1} and ϵ_{n+1} will both become as small as we please.

As in the case of the differential equation of the first order it is not worth while, as a rule, to investigate the convergence for the purpose of finding a sufficiently close approximation by graphical methods. It is better at once to tackle the task of drawing the approximations and to repeat the operations until no further improvement is obtained. The curve will then satisfy the differential equation as far as the graphical methods allow it to be recognized.

When the values of $f(x, y, z)$ or $g(x, y, z)$ become too large we can have recourse to the same device that we found useful with the differential equation of the first order. Instead of x, one of the other two variables y or z may be considered as independent, so that the equations take the form

$$\frac{dx}{dy} = \frac{1}{f(x, y, z)}, \quad \frac{dz}{dy} = \frac{g(x, y, z)}{f(x, y, z)},$$

or

$$\frac{dx}{dz} = \frac{1}{g(x, y, z)}, \quad \frac{dy}{dz} = \frac{f(x, y, z)}{g(x, y, z)},$$

or we may introduce a new system of coördinates x', y', z' and consider the resulting differential equations.

The second method for the integration of differential equations of the first order can also be generalized to include the second order. Let us again consider the more general case

$$\frac{dy}{dx} = f(x, y, z), \quad \frac{dz}{dx} = g(x, y, z).$$

Starting from a point x, y, z the changes of y and z (denoted by

k and l) can be calculated for a small change h of x by the following formulas analogous to those used for one differential equation of the first order:

$$k_1 = f(x, y, z)h; \qquad\qquad l_1 = g(x, y, z)h;$$

$$k_2 = f\left(x + \frac{h}{2},\ y + \frac{k_1}{2},\ z + \frac{l_1}{2}\right)h; \quad l_2 = g\left(x + \frac{h}{2},\ y + \frac{k_1}{2},\ z + \frac{l_1}{2}\right)h;$$

$$k_3 = f\left(x + \frac{h}{2},\ y + \frac{k_2}{2},\ z + \frac{l_2}{2}\right)h; \quad l_3 = g\left(x + \frac{h}{2},\ y + \frac{k_2}{2},\ z + \frac{l_2}{2}\right)h;$$

$$k_4 = f(x + h,\ y + k_3,\ z + l_3)h; \quad l_4 = g(x + h,\ y + k_3,\ z + l_3)h;$$

$$p = \frac{k_2 + k_3}{2},\ q = \frac{k_1 + k_4}{2}; \qquad p' = \frac{l_2 + l_3}{2},\ q' = \frac{l_1 + l_4}{2};$$

and with a high degree of approximation,

$$k = p + \tfrac{1}{3}(q - p); \quad l = p' + \tfrac{1}{3}(q' - p').$$

These calculations may be performed graphically. For this purpose the functions $f(x, y, z)$ and $g(x, y, z)$ must be given in some handy form. We notice that in our formulas the first argument assumes the values x, $x + h/2$, $x + h$. In the next step where $x + h$, $y + k$, $z + l$ are the coördinates of the starting point that play the same part that x, y, z played in the first step, we are free to make the change of the first argument the same as in the first step, so that in the formulas of the second step it assumes the values $x + h$, $x + \tfrac{3}{2}h$, $x + 2h$ and so on for the following steps. All the values of the first argument can thus be assumed equidistant. Let us denote these equidistant values by

$$x_0, x_1, x_2, x_3, \cdots.$$

The values of $f(x, y, z)$ and $g(x, y, z)$ appear in all our formulas only for the constant values

$$x = x_0, x_1, x_2, \cdots.$$

For each of these constants f and g are functions of two independent variables and as such may be represented graphically by drawings giving the curves $f =$ const. and $g =$ const., each value of x corresponding to a separate drawing. These drawings we must consider as the graphical form in which the differential equations are given. It may of course sometimes be very tiresome to translate the analytical form of a differential equation into a graphical form, but this trouble ought not to be laid to the account of the graphical method.

The method now is similar to that used for the differential equation of the first order. y and z are plotted as ordinates in the same system in which x is the abscissa. Equidistant parallels to the axis of ordinates are drawn

$$x = x_0, \; x = x_1, \; x = x_2, \; \text{etc.}$$

On the first $x = x_0$ we mark two points with ordinates y_0 and z_0, and from the drawing that gives the values of $f(x_0, \, y, \, z)$ and

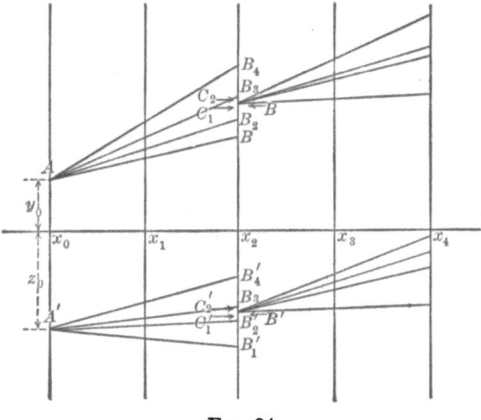

FIG. 94.

$g(x_0, \, y, \, z)$ as functions of y and z we read the values $f(x_0, \, y_0, \, z_0)$ and $g(x_0, \, y_0, \, z_0)$ and draw the lines from $x_0, \, y_0,$ and $x_0, \, z_0$ to the points

$$x_2, \, y_0 + k_1 \quad \text{and} \quad x_2, \, z_0 + l_1.$$

The intersections of these lines with the parallel $x = x_1$ furnishes the points

11

$$x_1, \ y_0 + \frac{k_1}{2} \quad \text{and} \quad x_1, \ z_0 + \frac{l_1}{2}.$$

With these ordinates we find from the second drawing the values

$$f\left(x_1, \ y_0 + \frac{k_1}{2}, \ z_0 + \frac{l_1}{2}\right) \quad \text{and} \quad g\left(x_1, \ y_0 + \frac{k_1}{2}, \ z_0 + \frac{l_1}{2}\right),$$

and by their help we can draw the lines from x_0, y_0 and x_0, z_0 to the points

$$x_2, \ y_0 + k_2 \quad \text{and} \quad x_2, \ y_0 + l_2.$$

The intersections of these lines with the line $x = x_1$ furnishes the points

$$x_1, \ y_0 + \frac{k_2}{2} \quad \text{and} \quad x_1, \ z_0 + \frac{l_2}{2},$$

and with these ordinates we find the values

$$f\left(x_1, \ y_0 + \frac{k_2}{2}, \ z_0 + \frac{l_2}{2}\right), \quad g\left(x_1, \ y_0 + \frac{k_2}{2}, \ z_0 + \frac{l_2}{2}\right),$$

which enable us to draw the lines from x_0, y_0 and x_0, z_0 to x_2, $y_0 + k_3$ and x_2, $z_0 + l_3$.

With these two ordinates we find from the third diagram ($x = x_2$) the values

$$f(x_2, \ y_0 + k_3, \ z_0 + l_3) \quad \text{and} \quad g(x_2, \ y_0 + k_3, \ z_0 + l_3),$$

which finally enable us to draw the lines from $x_0 y_0$ and $x_0 z_0$ to x_2, $y_0 + k_4$ and x_2, $z_0 + l_4$.

On the vertical line $x = x_2$ we thus obtain four points, B_1, B_2, B_3, B_4, corresponding to $y_0 + k_1$, $y_0 + k_2$, $y_0 + k_3$, $y_0 + k_4$ and four points, B_1', B_2', B_3', B_4', corresponding to $z_0 + l_1$, $z_0 + l_2$, $z_0 + l_3$, $z_0 + l_4$ (Fig. 94).

$B_2 B_3$ and $B_1 B_4$ are bisected by the points C_1 and C_2; $B_2' B_3'$ and $B_1' B_4'$ by the points C_1', C_2'. Finally $C_1 C_2$ and $C_1' C_2'$ are divided into three equal parts and the points B and B' are found in the dividing points nearest to C_1 and C_1'.

The same construction is then repeated with B and B' as starting points and furnishes two new points on the vertical

$x = x_4$ and so on. To test the accuracy the construction is repeated with intervals of x of double the size. The difference in the values of y and of z found for $x = x_4$ enables us to estimate the errors of the first construction—they are about one-fifteenth of the observed differences.

Both methods are without difficulty generalized for the integration of differential equations of any order. We can write a differential equation of the nth order in the form

$$\frac{d^n x}{dt^n} = f\left(t, x, \frac{dx}{dt}, \cdots, \frac{dx^{n-1}}{dt^{n-1}}\right),$$

or in the form of n simultaneous equations of the first order

$$\frac{dx}{dt} = x_1,$$

$$\frac{dx_1}{dt} = x_2,$$

$$\cdot\ \cdot\ \cdot\ \cdot\ \cdot\ \cdot$$

$$\frac{dx_{n-2}}{dt} = x_{n-1},$$

$$\frac{dx_{n-1}}{dt} = f(t, x, x_1, x_2, \cdots x_{n-1}).$$

A more general and more symmetrical form is

$$\frac{dx}{dt} = f_1(t, x, x_1, \cdots x_{n-1}),$$

$$\frac{dx_1}{dt} = f_2(t, x, x_1, \cdots x_{n-1}),$$

$$\cdot\ \cdot\ \cdot\ \cdot\ \cdot\ \cdot\ \cdot\ \cdot\ \cdot\ \cdot\ \cdot$$

$$\frac{dx_{n-1}}{dt} = f_n(t, x, x_1, \cdots x_{n-1}).$$

The functions $x, x_1, x_2, \cdots x_{n-1}$ are then represented as ordinates to the abscissa t, so that we have n different curves. When the function $f(t, x, x_1, x_2, \cdots x_{n-1})$ is given in a handy form, so that its value may be quickly found for any given values of t, x, x_1, $\cdots x_{n-1}$, there is no difficulty in constructing n curves whose ordinates represent the functions x, x_1, x_2, $\cdots x_{n-1}$. Starting from given values of t, x, x_1, x_2, $\cdots x_{n-1}$ we have only to apply the same methods that have been explained for the first and the second order.

COLUMBIA UNIVERSITY PRESS

Columbia University in the City of New York

COLUMBIA UNIVERSITY LECTURES

ADAMS LECTURES

Graphical Methods. By CARL RUNGE, Ph.D., Professor of Applied Mathematics in the University of Göttingen ; Kaiser Wilhelm Professor of German History and Institutions for the year 1909–1910. 8vo, cloth, pp. ix+148. Price, $1.50 *net.*

JULIUS BEER LECTURES

Social Evolution and Political Theory. By LEONARD T. HOBHOUSE, Professor of Sociology in the University of London. 12mo, cloth, pp. ix+218. Price, $1.50 *et.*

BLUMENTHAL LECTURES

Political Problems of American Development. By ALBERT SHAW, LL.D., Editor of the *Review of Reviews*. 12mo, cloth, pp. vii+268. Price, $1.50 *net.*

Constitutional Government in the United States. By WOODROW WILSON, LL.D., President of Princeton University. 12mo, cloth, pp. vii+236. Price, $1.50 *net.*

The Principles of Politics from the Viewpoint of the American Citizen. By JEREMIAH W. JENKS, LL.D., Professor of Political Economy and Politics in Cornell University. 12mo, cloth, pp. xviii+187. Price, $1.50 *net.*

The Cost of Our National Government. By HENRY JONES FORD, Professor of Politics in Princeton University. 12mo, cloth, pp. xv+147. Price, $1.50 *net.*

The Business of Congress. By HON. SAMUEL W. McCALL, Member of Congress for Massachusetts. 12mo, cloth, pp. vii+215. Price, $1.50 *net.*

CARPENTIER LECTURES

The Nature and Sources of the Law. By JOHN CHIPMAN GRAY, LL.D., Royall Professor of Law in Harvard University. 12mo, cloth, pp. xii+332. Price, $1.50 *net.*

World Organization as Affected by the Nature of the Modern State. By HON. DAVID JAYNE HILL, American Ambassador to Germany. 12mo, cloth, pp. ix+214. Price, $1.50 *net.*

The Genius of the Common Law. By the RT. HON. SIR FREDERICK POLLOCK, Bart., D.C.L., LL.D., Bencher of Lincoln's Inn, Barrister-at-Law. 12mo, cloth, pp. vii+141. Price, $1.50 *net.*

LEMCKE & BUECHNER, Agents

30-32 West 27th Street, New York

Bei Fragen zur Produktsicherheit wenden Sie sich bitte an:
If you have any questions regarding product safety,
please contact:

Walter de Gruyter GmbH
Genthiner Straße 13
10785 Berlin
productsafety@degruyterbrill.com